TAIWAN'S DEMOCRATIZATION

Forces Behind the New Momentum

STUDIES ON CONTEMPORARY TAIWAN

The Studies on Contemporary Taiwan Series features
academic works by Chinese scholars from Taiwan on the
island's key political, economic, social, and cultural issues.
The works for the series were selected from outstanding
doctoral research produced by Taiwanese scholars while
pursuing degrees at overseas universities.
In what ways does the China reunification issue affect
Taiwan? How has increased democratization affected
policymaking in Taiwan? What are the prospects for further
democratization in Taiwan? The authors, having studied
abroad, bring a unique international perspective to these and
other critical questions surrounding their own nation's
development and future prospects.
The books in this series also have benefited from the review
and input of distinguished members of the international
community of Taiwan scholars, who have helped the authors
and editors to refine and develop the research for publication.

TAIWAN'S DEMOCRATIZATION

Forces Behind the New Momentum

Jaushieh Joseph Wu

HONG KONG
OXFORD UNIVERSITY PRESS
OXFORD NEW YORK
1995

Oxford University Press
Oxford New York
Athens Auckland Bangkok Bombay
Calcutta Cape Town Dar es Salaam Delhi
Florence Hong Kong Istanbul Karachi
Kuala Lumpur Madras Madrid Melbourne
Mexico City Nairobi Paris Singapore
Taipei Tokyo Toronto
and associated companies in
Berlin Ibadan

Oxford is a trade mark of Oxford University Press

First published 1995

Published in the United States
by Oxford University Press, New York

© Oxford University Press 1995

The 'Studies on Contemporary Taiwan' series is published in cooperation with
Sinorama Magazine, Taipei, Taiwan, Republic of China

British Library Cataloguing in Publication Data available
Library of Congress Cataloging-in-Publication Data
Wu, Jaushieh Joseph, 1954–
Taiwan's democratization : forces behind the new momentum /
Jaushieh Joseph Wu.
p.   cm. — (Studies on contemporary Taiwan)
Includes bibliographical references and index.
ISBN 0–19–586499–9
1. Democracy — Taiwan.   2. Taiwan — Politics and government — 1988–
3. Political participation — Taiwan.   I. Title.   II. Series.
JQ1536.W84   1995
320.95124'9'09048—dc20   94–31422
                                                      CIP

Printed in Hong Kong
Published by Oxford University Press (Hong Kong) Ltd
18/F Warwick House, Taikoo Place, 979 King's Road, Quarry Bay, Hong Kong

To Su Ru-yuh and Wu Dee

Preface

WRITING on a topic which is still an ongoing process is not an easy undertaking. While I have been writing this book, Taiwan has undergone two constitutional revisions, three important elections, and a change of electoral system. In addition, the factions of the opposition party, brought together in a fragile coalition, have called at least a temporary truce in the hope that they can bring down the government. The ruling party, on the other hand, is embroiled in a very serious internal power struggle. The Taiwan government has also revised its stringent 'one China' policy and applied to enter the United Nations, though Beijing has reiterated in a recent white paper its own tough stance that Taiwan is a part of China. Only a short time ago, President Lee Teng-hui seemed secure in his position, with a very high popularity rating among Taiwan's voters. But since the Kuomintang's Fourteenth Congress his leadership has been challenged by the party's own factions. The seemingly unshakable ruling position of the KMT is also threatened by the party's difficulties in maintaining a majority in the legislature. Changes like these may come as no surprise to the average citizen accustomed to Taiwan politics, but they are viewed with great excitement by academics, who rarely have a chance to witness a new polity in the making. However, these rapid and unpredictable changes make it very difficult to write a book on Taiwan's politics that will not be out of date as soon as it is published.

Yet despite these drawbacks, looking back at the process of democratic transition from different perspectives may allow those who are interested in political democratization in general and the Taiwan case in particular to gain a broader and more systematic understanding of the forces behind the change and the factors affecting the transition. A whole new generation of political scientists in Taiwan's universities have begun the great undertaking of studying their own country, and a substantial amount of research has already been published. Hopefully, more books like this one will be on the shelves soon.

I have a number of people to thank for help with this book.

First of all, I am deeply grateful to those who granted opportunities for personal interviews. In addition, professors John Hsieh, Chen I-yen, Hwang Teh-fu, and Liu I-chou at the Election Studies Center, National Chengchi University, have provided crucial election and survey data. Dr Szu-yin Ho, a colleague and a very good friend at the Institute of International Relations (IIR) and himself a brilliant political scientist, has been very generous in providing insights at every stage of the process. Ms Judith Fletcher, another colleague at the IIR, has spent long hours reviewing the entire manuscript and attempting to clear up all vagueness and inconsistencies. Ms SueEllen Chen, a young free spirit at the IIR, has typed up part of the manuscript and organized the bibliography for me. These friends and colleagues truly deserve recognition and appreciation.

JAUSHIEH JOSEPH WU
Mucha, Taipei
Taiwan

Contents

Tables

Figures

Chapter 1

Introduction: Studying Taiwan's Democratization

WHILE popping a champagne cork in celebration of his party's election success on the night of 19 December 1992, Hsu Hsin-liang, chairman of the six-year-old Democratic Progressive Party (DPP), proclaimed that 'two-party politics are firmly consolidated in Taiwan' (*United Daily*, 20 December 1992: 2). Indeed, the extraordinary performance of the opposition DPP in the Legislative Yuan election (35 per cent of the popular vote and 30 per cent of the total seats) was a milestone in the progress towards democracy in Taiwan which started in the mid-1980s. The ruling Nationalist Party (Kuomintang, KMT) no longer exclusively dominates politics and society, as it had done since 1949 when the government of the Republic of China (ROC) was defeated by the Chinese communists and retreated to Taiwan. Despite this electoral setback, the KMT remains the majority party in Taiwan. Nevertheless, under a strong challenge from the opposition and public pressure for further reform, the government is gearing up for transition to full democracy.

The relatively peaceful evolution from authoritarian rule in recent years has mirrored Taiwan's economic performance over the past four decades. Although Taiwanese themselves are still debating how the island's economy can be further improved, few would question that Taiwan's economic development has been no small miracle. A very consistent average annual growth rate of about 7 per cent, plus a per capita GNP of US$10,202 for 1992 and US$10,566 for 1993, has ranked Taiwan among the world's fastest growing economies (Directorate-General of Budget, Accounting, and Statistics [BAS], 1994: 4). This economic success has enabled the government to allocate a significant portion of its financial resources to infrastructure, research and development, education, and other public projects. A whole range of wealth redistributive measures have also given the island a remarkably equitable income distribution. These achievements have led scholars of Third World development to view Taiwan as a model for developing countries.

As far as political development is concerned, Taiwan has only begun to 'take off' since 1986, when opposition forces formed themselves into a technically illegal political party, the DPP. The following year, martial law was abolished, opening the door to press freedom and freedom of speech. The two main chambers of the parliament (the Legislative Yuan and the National Assembly) were revamped at the end of 1991 when aging deputies who had been maintained in office since their election on the mainland in the late 1940s were forced to retire. In 1992, the constitution was amended in an attempt to create a more effective government. Within just a few years, Taiwan's KMT-dominated authoritarian party-state structure was dismantled; what is emerging to take its place is a fairly competitive and open political system. According to accepted definitions, Taiwan is not yet a full democracy—the political system is still wavering between parliamentarianism and presidentialism, for example, and the electronic media are still controlled by the ruling party—but with the government under strong public pressure to resolve these outstanding issues, democracy is surely a none too distant prospect. This relatively peaceful political transition has brought Taiwan in line with the worldwide trend of democratization and has also attracted the attention of students of Third World political development.

Indeed, Taiwan's democratization has not been carried out in isolation. Beginning with the successful transition to democratic government in Spain and Portugal, followed in Greece and Turkey, continuing with the movement away from military dictatorship in several Latin American nations, and the popular uprisings against dictatorial rule in East Asia, the world has experienced what Huntington (1991) has called a 'third wave' of democratization. Most notable in this regard are the totally unexpected changes in the states of the former Soviet Union and Eastern Europe, most of which are at least attempting to establish democratic systems.

Although Taiwan is a part of this wave of democratization, it is quite unique in several respects. Unlike South Korea and the Philippines or the Latin American countries, Taiwan did not have any previous experience of democracy, though the KMT, for all its Leninist party structure and penetration into society, does lay claim to a democratic ideology. After the death of strongman Chiang Ching-kuo in early 1988, the ruling party did not collapse, nor was it replaced by a new ruling élite, as happened in Spain after Franco. Moreover, the KMT has managed to win large majorities in all the key elections held since the transition process began. The most noticeable distinction in Taiwan's case, though, lies in its historical confrontation with mainland China, and the situation in which the governments in Taipei and Beijing continue to claim sovereignty over each other's territory. Taiwan also faces an internal division over whether it should eventually become part of a reunified China. Taiwan's very distinctiveness, however, may help students of democratization to better understand the complexity of the phenomenon.

As a basis for their theories of democratization political scientists have been attempting for a long time to define democracy. Some prominent examples are Lipset (1959), Dahl (1971), Linz (1975), Powell (1982), Lijphart (1984), and O'Donnell et. al. (1986), to name just a few. Common to all these definitions is a belief that the essential ingredients of a democratic political system are basic freedom and fair and free electoral competition for key government decision-making positions. More detailed qualifications for democracy might include elements such as universal suffrage, a free flow of information, majority rule, and equality.

Evaluating Taiwan's current status according to these yardsticks, one finds that with the exception of the election of key decision-makers (president or premier), Taiwan largely makes the grade as a democracy, and according to the definitions of Higley and Gunther (1992: 20–4), and Diamond, Linz, and Lipset (1989: 45–7), appears to be in the process of consolidating its democracy. The government guarantees civil liberties, and free and open elections have been held for the Legislative Yuan, the National Assembly, and other sub-national government offices. Observers from around the world judged the elections in December 1992 to be democratic in nature, although somewhat unfair because of government domination of the electronic media.[1] In the area of institutionalization, no decision has yet been made on how the government's top decision-maker should be elected, but according to Article 12 of the 1991 addenda to the constitution, the method of electing the president will be decided before 1995 by an extraordinary session of the National Assembly. If and when that decision is taken, Taiwan will truly have consolidated its democracy. Taiwan's polity today stands in stark contrast to the era of 'white terror' and the KMT's authoritarian rule in the 1950s through the 1970s. To have achieved all this without significant disruption of the political and economic processes is indeed remarkable.

Democracy and democratization is one of the main topics, if not the key topic, of comparative political studies and studies of political development. Although the attention of scholars shifted temporarily in the 1970s and the first part of 1980s to the problems of dependency and underdevelopment, since the 'third wave' of democratization got well underway, a significant number of comparative studies have appeared which have provoked political scientists to question old theories and assumptions about democratization and to look more closely at each important case of democratic transition. Among the most notable efforts in this regard are those by O'Donnell, Schmitter, and Whitehead (1986), Diamond, Linz, and Lipset (1989), Cheng (1992), and Higley and Gunther (1992).

However, despite its importance as a case of peaceful transition from authoritarian rule, Taiwan has only recently begun to attract serious attention from political scientists. The relative lack of in-depth studies comes as no surprise considering the fact that Taiwan has usually been treated as an adjunct to Chinese studies. Few scholars specialize in Taiwan, and 'Taiwan studies' as a subject is virtually non-existent in universities. Consequently, students of Taiwan's political development must rely on the work of a few scholars who publish in political science and Asian studies journals. There is no doubt, though, that a closer look at the Taiwan case would shed more light on the topic of democratization, as it would offer political scientists an opportunity to re-examine the process of democratic development in a different political, economic, international, and cultural setting.

The comparative studies mentioned above do offer several important perspectives on the Taiwan case that focus on its most significant aspects while remaining within a comparative framework. The first perspective is the level of socioeconomic development and its association with democracy (Deutch, 1953; Lerner, 1958; Lipset, 1959; Almond and Verba, 1963;

Cutright, 1969; Huntington, 1984). The main argument of these studies is that socioeconomic development, including the availability of education, access to mass communications, and transportation, is conducive to democratization. In the case of Taiwan, then, one must ask whether the island has attained a sufficiently high level of socioeconomic development to allow its population to participate meaningfully in politics, particularly in elections, so as to ensure the consolidation and continuation of a democratic political system. The answer to this question is that Taiwan has far surpassed the basic level of economic development required for the development of democracy (see Diamond, 1992b).

The second perspective is the consolidation and institutionalization of democracy (Higley and Gunther, 1992; Diamond, Linz, and Lipset, 1989). Elections and the main participating groups, particularly political parties vying for power, are the most important elements in the building of democratic institutions. Analyses of Taiwan's transition to democracy should include discussions on Taiwan's main political parties and their comparative performance at the polls. In Taiwan, one should look in particular at the 1992 election for the Legislative Yuan, the first in which all the seats were contested. It is the results of this election that should be used to demonstrate the basis of support of the two main political parties, and possibly to predict Taiwan's political future.

The third perspective is the élite calculation of repression versus toleration and élite settlement (Higley and Gunther, 1992; Huntington, 1992; O'Donnell, Schmitter, and Whitehead, 1986). As argued by Higley and Gunther (1992), leaders of main political groups and key decision-makers play a crucial role in deciding the course of political change. A decision by the ruling élites either to repress or to tolerate opposition, and that by opposition élites to rebel or cooperate, can determine the process and the outcome of democratic transition. In order to gauge the attitudes of élite groups towards the building of a democratic political system, an in-depth discussion of their political views is needed. An understanding of the complexity of the problems facing Taiwan's further democratization requires the identification of key issues confronting the consolidation of democracy. The views of élite groups may then be classified and mapped.

For the first two perspectives, a review of the literature

and a re-examination of news reports may yield substantial material. However, a more systematic approach is necessary in order to gauge the attitudes of political élites. The author has therefore conducted intensive interviews of more than twenty political leaders of national importance between 1988 and 1993. The views expressed by these leaders on key political issues confronting Taiwan will be the main source for evaluating the relationship between political transition and élite settlement. The names of the interviewees are listed in Appendix 1.

As stated above, the main purpose of this study is to take a close look at the process of Taiwan's recent democratization, particularly the factors driving the transition process and certain issues of national concern which might delay the achievement of full democracy. The book will begin with a brief account of the Chinese people's quest for democracy and the historical background of Taiwan. This chapter will then outline the situation in Taiwan prior to KMT rule and finally deal more extensively with the characteristics of the regime at different periods since 1945.

Chapter 3 will examine the move away from authoritarian rule, including Chiang Ching-kuo's decision to liberalize the polity and the role of Taiwan's socioeconomic development. Most studies indicate that Taiwan has achieved both rapid economic growth and a remarkably equitable distribution of wealth, two seemingly contradictory goals for most Third World countries. Moreover, the population in general is well-educated, urbanized, and has ready access to the mass media. A substantial number are willing to participate in protest activities to make their demands known to the government.

Chapter 4 looks in detail at Taiwan's political élites. A comparison of the electoral strengths and weaknesses of the two main political parties in Taiwan—the KMT and the DPP—will be followed by a discussion of the social backgrounds of each of the élite groups and the social bases of élite support. It will be seen that both the KMT and the DPP are catch-all parties which are very careful to avoid recruiting support from specific social groups.

Chapter 5 will provide a detailed description of the political élite's views on such issues as parliamentary reform, amending the constitution, and the unification/independence controversy. On each of these issues, the ruling group can be clearly divided into liberal and conservative camps, and the opposition into radicals and moderates. Each camp's views will be outlined for the purpose of comparison. Interview data will be the main source of evidence in this chapter, and the conclusion drawn is that there is a fundamental conflict among the élite regarding how democratization can be achieved.

The main finding in the concluding chapter is that all the socioeconomic prerequisites for democracy have generally been met in Taiwan, and the country is on the path to a fully democratic system. The main difficulties in the future lie in the conflicting views of the national political élites regarding how important issues should be resolved and how democracy can be achieved. Finally, the author will offer his own suggestions for resolving this conflict.

Chapter 2

∎

Democratization in Perspective

IT is not the purpose here to write a comprehensive contemporary history of Taiwan. Rather, this chapter will focus only on those historical events that have affected Taiwan's political and socioeconomic development in recent decades. Democratization is not an isolated event in Taiwan's history; it is the culmination of an historical process. The political thinking of the May Fourth Movement, the economic infrastructure developed by the Japanese colonial administration, Sun Yat-sen's grand constitutional design, Chiang Kai-shek's Leninist-style regime, and above all, Taiwan's peculiar relationship with China: all these have either acted as impetuses or obstacles in Taiwan's quest for democracy. The following discussion will be limited to those aspects of China's and Taiwan's history that have had an impact on the democratic development of the 1980s and 1990s.

The Quest for Democracy in Contemporary China

Relations between Taiwan and China and Taiwan's identity as a part of China are still issues of intense debate among Chinese on both sides of the Taiwan Strait. A few words in this chapter can hardly resolve a controversy which involves legal, international, political, cultural, and historical factors on both sides. One real danger in any academic discussion of this issue is that one may provoke criticism from either Chinese or Taiwanese nationalists and be branded a traitor by one side or the other. A more practical and less dangerous approach would be to look at the historical ties between the island and the mainland, particularly the Nationalist government's roots in China. From this angle, Taiwan's democratization process can be understood as part of a continuing effort by the Chinese people to create a democratic political system.

Events such as Sun Yat-sen's Nationalist Revolution of 1911 and the May Fourth Movement of 1919 introduced the idea of democracy to Chinese society, an idea that was kept alive throughout the period of authoritarian rule that followed. Even though these early movements produced no lasting democratic experiments, they do help us to understand the sudden surge in the democratic movement in Taiwan in the 1980s and 1990s. In the same way, the success of the recent democratization drive also gives new meaning to China's past history.

There have been several serious attempts to modernize China in recent history, and two of these were aimed specifically at building a democratic China. The first effort to modernize the country occurred in the 1860s and was sparked off by the incursions of the Western powers, whose success rested on their military might and modern technology. The reformers sought to utilize the material achievements of the West without upsetting the traditional hierarchy. The failure of this attempt prompted some people, particularly intellectuals, to look beyond the material aspects of Western culture. After China's defeat in the Sino-Japanese War of 1895, various provinces attempted to set-up assemblies and to force corrupt and incompetent leaders to incorporate new ideas from below. However, the tempo of change was not fast enough for some, who wanted to establish an entirely new polity right away. The 1911 revolution, initiated by Sun Yat-sen, was intended not just to overthrow the Manchu dynasty, but to pave the way for an extensive social, economic, and political renewal. By 1919, however, the aspirations of the revolutionaries had still not been fulfilled. Warlords were in control of much of China and the Allies had betrayed the Chinese by agreeing to Japanese demands at the Paris Peace Conference. This paved the way for the May Fourth Movement, which started as a series of demonstrations against the Peking government's concessions to the Japanese but was actually a broad movement of cultural and intellectual renewal.

Taiwan's recent democratization is closely related to both the 1911 Revolution and the May Fourth Movement, and these two events will be examined in detail so that some controversial issues involved in Taiwan's transition process can be set in historical perspective.

Sun Yat-sen's Revolution

The 1911 revolution was the first attempt by the Chinese to build a democratic political system. Sun Yat-sen, leader of the revolution and founding father of the Republic of China, set out his political creed in a series of lectures on the Three Principles of the People (*Sanmin zhuyi*)—*minzu* (nationalism), *minquan* (democracy), and *minsheng* (people's livelihood)—which formed the ideological tenets of the revolution and the newly established republic. Of these three principles, *minquan* is literally translated as 'people's sovereignty' and more broadly interpreted as 'democracy'. Indeed, in his speeches, Sun clearly attributed this meaning to the term: 'Briefly, government is a thing of the people and by the people; it is control of the affairs of all the people. The power of control is political sovereignty, and where the people control the government we speak of the "people's sovereignty"' (Sun, 1985: 52). Later in the same series of lectures he explains how this sovereignty will be exercised:

> When democracy is highly developed and methods of controlling government are perfected, the government will have great power, but the people will only have to make their opinions known in their national congress. . . . We must have a complete and powerful government organ, and at the same time have a compact method of popular sovereignty to exercise control over the government organ. (Sun, 1985: 134, 141)

> Only when the people have these four powers [election, recall, initiative, and veto] can we say that there is a full measure of democracy, and only where these four powers are effectively applied can we say that there is thoroughgoing, direct, popular sovereignty. Before there was any complete democracy, people elected their officials and representatives and then could not hold them responsible. This was only indirect democracy or a representative system of government. The people could not control the government directly but only through their representatives. For direct control of the government it is necessary that the people practice these four forms of popular sovereignty. Only then can we speak of government by all the people. (Sun, 1985: 143–4)

Although it is quite obvious that Sun wanted to create a democratic China, the kind of democracy he envisioned was not entirely based upon the democracy in practice in the West. Instead, it has its origins in the system of checks and balances advocated by Montesquieu. It is not the kind of checks and balances among government institutions found in the American system of government, but a checks-and-balance system between the people and the government.

In order to establish a political system in which the government has power and the people have sovereignty, Sun also devised a central government structure unique to the Republic of China. According to Sun's blueprint, the government would consist of five branches or Yuan: Executive, Legislative, Judicial, Examination, and Control. The Executive, Legislative, and Judicial branches corresponded with Montesquieu's ideas and the Western tradition of the separation of powers. The Examination and Control Yuan had their origins in the government structure of imperial China. The former was to be in charge of recruiting civil servants through competitive examinations and the latter was to act as a guard against corruption. These five branches together formed the 'political power' of the government.

'Popular sovereignty' was to be embodied in the National Assembly, which was designed to exercise the rights of election and recall of government leaders, initiative, and veto. Although Sun's ideal was Swiss-style 'direct democracy', he justified this apparent shortcoming by arguing that China was so large and populous that direct democracy was impossible. According to Sun, the essence of direct democracy would be preserved in the National Assembly which would directly represent the people in the central government.

Despite the failure of the 1911 revolution to produce a democratic political system for the Chinese people, this governmental set-up was largely preserved by the 1947 constitution drafted by the Nationalist government in Nanjing. The constitution remained in force in Taiwan after the Nationalists were forced to flee to the island in 1949. It was not amended until 1991, and its suitability gradually became one of the most controversial issues in Taiwan's democratic transition.

In addition to adopting Sun's design for the central government, the ruling KMT has also retained the Three Principles of the People as Taiwan's official ideology. Before the

government-initiated liberalization of the mid-1980s, the slogan 'Liberate China with the Three Principles of the People'
(Sanmin zhuyi jiejiu Zhongguo) was to be seen everywhere
on public buildings and it was widely cited by Nationalist
officials in their political speeches. The slogan was later
changed to 'Reunify China with the principles of freedom,
democracy, and prosperity' (Ziyou minzhu junfu tongyi Zhongguo), which comes to the same thing but does not carry such
heavy ideological connotations. The Three Principles are also
required courses in high schools and colleges and are one
of the subjects tested in the college entrance examination.
Ideological indoctrination was furthered by the establishment
of graduate institutes of the Three Principles in several major
universities to train teachers and propagandists. Anybody with
at least a high-school education in Taiwan is well-informed
about Sun's ideology; consequently, his influence goes far
beyond institutional aspects.

Moreover, today's Kuomintang is of course the same party
that Sun founded and the same one that overthrew the Qing
dynasty. The party has retained this prestige in spite of its
defeat by the communists in 1949 and its subsequent retreat
to Taiwan. For many years, the KMT's official line was that
the Nationalist government was the only legitimate representative of the Chinese people and that the communists were
nothing but bandits. This belief led to the preservation of the
original constitution, the declaration of an emergency decree,[1]
and the freezing in office of the pre-1949 parliament. These
issues became focuses of controversy between reformers and
conservatives during Taiwan's democratization process.

Although Sun's constitutional set-up has been revised to
some extent to suit the new political conditions in Taiwan,
it has given Taiwan a strong central government administration and a powerful national bureaucracy. The ability of the
government to initiate effective economic policies has allowed
Taiwan to attain a fairly high level of socioeconomic development, which has proved to be very important in pressur-
ing the government to relax its rigid political control. Moreover,
the KMT's very loyalty to Sun's doctrine of popular sovereignty, not to mention constitutional guarantees of civil liberties, freedom of speech, and regular democratic elections,
contained the seeds of its eventual liberalization. In sum, Sun's
revolution and the democratic ideas advocated by him formed

a legacy that people could utilize to challenge the authoritarian rule of the KMT government and which contributed
to the rise of the democracy movements of the 1980s.

May Fourth Movement

The May Fourth Movement, which takes its name from a series
of mass demonstrations and strikes against China's treatment
at the Versailles Conference, was the first serious attempt by
young intellectuals to tackle the cultural and ideological aspects
of a social revolution. These intellectuals later divided into
two factions—one advocating Bolshevik-style socialism, many
of whose members went on to become the founding fathers
of the People's Republic of China (PRC), and the other tend-
ing towards Western liberalism and eventually throwing in its
lot with the Nationalist government. The movement had a
significant impact on the ideological development of liberal
intellectuals in Taiwan, where the earliest democratic movement was initiated by May Fourth intellectuals.

In the years prior to these demonstrations, a loose coalition
of intellectuals educated overseas, including such famous
names as Hu Shi and Chen Duxiu, formed around the campus
of Peking University and the magazine *New Youth* (*Xin qing-
nian*). They were searching for ways to modernize China,
which was weak, disorganized, and helpless in the face of the
aggression of the industrialized powers. In particular, these intellectuals addressed the nation about the importance of democracy and science to the modernization of China.

No one has stated the importance of the May Fourth Movement
better than Lucien Bianco, the French historian of China:

> What was the May Fourth Movement? In part it was an intel-
> lectual reawakening, sometimes called the Chinese Renaissance
> but in fact, if a comparison with European intellectual his-
> tory must be made, closer in spirit to the eighteenth century.
> The May Fourth Movement was a kind of Chinese Enlightenment,
> a movement that advanced such eminently reasonable ideals
> as science and democracy. More important, it was a ground-
> clearing enterprise; it foreshadowed and paved the way for
> 1949 just as Voltaire had for 1789. The established order to
> be crushed in the Chinese case was also the Church, or at
> least that religion without dogma, clergy, or prescribed form
> of worship known as Confucianism. . . . The May Fourth

Movement called into question the very basis of Chinese society. The young students who wished to be rid once and for all of the evils they denounced were not wrong to hurl themselves against the Confucian citadel crying 'Overthrow Confucius & Sons!' (Bianco, 1971: 27–28)

The literary revolution which was part of this movement brought about a dramatic change in the Chinese written language, bringing it much closer to the vernacular and thus opening the door to popular education. In addition, the spread of socialist and liberal ideas had a deep impact on both the Chinese Communist Party (CCP) and the KMT. Indeed, the most prominent intellectuals on both sides claim their heritage from the May Fourth Movement. Hu Shi, one of the most prominent May Fourth intellectuals, continued to play an important role in academic circles in Taiwan until his death. In addition to his intellectual role, Hu was respected by the Nationalist regime and once served as ambassador to the United States. He was also one of the initiators of the abortive attempt to found a Chinese Democratic Party in 1960.

From this period on, warlord fighting, Japanese invasion, and civil war followed one upon the other until 1949. Intellectuals of the period were faced with the dilemma whether to press for democracy or to support a strong central government that could unify China and repel the invaders. As the need to drive out the invaders became more urgent, the May Fourth Movement's intellectuals gradually opted to join either the communist or the Nationalist camp, both of which strongly emphasized the centralization of government power. In short, China was given no real chance to experiment with multi-party democracy.

What is more, the society in general was still basically agricultural, and the population by-and-large uneducated and isolated from the outside world. There was the lack of a modern infrastructure, a modern industrial economy, and a communications network, without which it was impossible to effect the mass participation in politics necessary for democracy. Without a mobilized population demanding more political freedom and democracy, intellectual movements could have no significant impact on the political system.

Nevertheless, the May Fourth Movement left an important legacy to intellectuals in both China and Taiwan. May Fourth leaders who went to Taiwan dominated intellectual life there for more than two decades. They remained loyal to the idea that only democracy and science could save China. The abortive attempt to form an opposition party in 1959–60 was organized and led by May Fourth intellectuals, including Hu Shi, Lei Chen, and Yin Hai-kuang, a respected professor at National Taiwan University.

Major opposition leaders in Taiwan today still emphasize their connection with the events of 1959–60. On the eve of the founding of the DPP in September 1986, such leading lights of the earlier movement as Fei Hsi-ping, a legislator, and Fu Cheng, a political science professor at Suzhou University, were among those who insisted that the word 'democratic' be included in the name of the new party. One of the chief initiators of the 1959–60 movement, Lei Chen, who was released from prison only after martial law was lifted, was honoured by the DPP as a pioneer of Taiwan's democracy movement. Thus the movement of the mid-1980s could trace its roots back to the events of 1959–60 and to the May Fourth spirit these events represented.

Taiwan: The People

Taiwan is an island off the southeast coast of mainland China, called '*Ilha Formosa*' (beautiful island) by Portuguese adventurers in the sixteenth century. It is separated from China by the Taiwan Strait, which is only about 160 kilometres wide at its narrowest point. The original inhabitants of the island are closer in ethnic origin to Malays and Indonesians than to the Han people of China. They consist of nine main tribes, each with its own distinctive language and cultural traits that are entirely different from those of the Chinese.

When Chinese immigrants started to occupy the fertile land along the western coast of Taiwan, the aborigines, who lived by fishing, hunting, and gathering, gradually retreated to the mountains, and under KMT rule came to be known as *shanbao*, 'mountain compatriots'. Through continuous immigration, the Chinese in Taiwan eventually outnumbered the aborigines. By the time the KMT government retreated to Taiwan in 1949, aborigines made up less than 2 per cent of the total population, and today they number approximately 330,000. Most of the Taiwanese aborigines live in the central mountain areas and eastern Taiwan. There is a striking similarity between their fate and that of the American Indians.

The Chinese population is today divided into two main groups: Taiwanese and mainlanders. Even though both groups are immigrants from mainland China and are ethnically Han people, the term 'mainlander' is usually applied to those who came to Taiwan with the Nationalist government in 1949 and including their descendants born in Taiwan. Approximately 10–15 per cent of the population can now be categorized as 'mainlanders' (Tsai, 1987: 4). When they were evacuated from China in 1949 they formed a heterogeneous group of about one million from different regions of the mainland, speaking a variety of dialects. Most of the mainlanders were either soldiers or civil servants and academics. These were the people who filled the political vacuum left by the departing Japanese occupiers, and in the early years of KMT rule the ruling class was composed almost exclusively of mainlanders.

Among the Taiwanese, there are two sub-groups, distinguished chiefly by their dialect: Fukienese (customarily called Taiwanese since 1949) and Hakka. The ancestors of the Fukienese came from southern Fujian Province and their dialect is spoken only south of the Min River. Hakka people are found all over southern China, though the ancestors of Taiwan's Hakka population migrated from Guangdong and Fujian provinces. Approximately 85 per cent of non-mainlanders can be categorized as Fukienese and about 15 per cent as Hakka (Sih et al., 1973: 362). It is generally the case that mainlanders live in urban areas and Hakka people are concentrated in the mountainous areas of Taoyuan, Miaoli, Taichung, and Pingtung counties. Communal conflicts between Fukienese and Hakka, and even between Fukienese from different areas of the province, were quite fierce during the earlier years of migration to the island. These conflicts most likely originated in mutual distrust arising from the inability to communicate. Since the nationalist government enforced compulsory education in Mandarin Chinese, however, such conflicts have become rare.

It was conflict between Taiwanese as a whole and mainlanders that created the biggest problem for the KMT during the early years of its regime on the island. When the Japanese troops in Taiwan surrendered to the Nationalists in 1945, most

Taiwanese rejoiced to see the end of Japanese rule and their return to China. However, this initial welcome soon turned sour as Taiwanese became better acquainted with the arriving mainlander soldiers and officials. After half a century of life under the Japanese, Taiwanese were highly disciplined and enjoyed a much higher standard of living than the mainlander troops. The mainlanders, they felt, were preoccupied with enriching themselves rather than managing the island's affairs. The sharp contrast between the mainlanders and the disciplined Japanese soon made the Taiwanese realize that 'returning to the motherland' was not such a joyful prospect (see Peng, 1973; Lee, 1991; Lin, 1990).

The Nationalist troops, for their part, were very suspicious of the islanders, many of whom behaved and dressed very much like Japanese, whom the mainlanders considered to be their bitterest enemies. Educated Taiwanese spoke fluent Japanese but not a word of Mandarin. Coming from the war-devastated mainland, the Nationalist troops and officials were easily corrupted in the relatively affluent environment of Taiwan (see Kerr, 1965; Lee, 1991; Lin, 1990).

The clash of the two cultures came to a head on 28 February 1947, when a Taiwanese woman selling untaxed cigarettes was beaten to death by Nationalist police. The incident soon developed into an island-wide revolt against Nationalist rule. The heavy-handed suppression that followed, carried out by Nationalist reinforcements sent from the mainland, claimed the lives of between 10,000 and 20,000 Taiwanese (Kerr, 1965: 310; Lee, 1991: 171–2). The suppression eliminated an entire generation of the Taiwanese élite, paving the way for Chiang Kai-shek's rule. The bloody massacre left a deep rift between Taiwanese and mainlanders, and the repercussions of what came to be known as the February 28 Incident are still affecting Taiwan politics today.

Discussion of the February 28 Incident was taboo for many years, most people refraining from talking about it for fear of government reprisals. However, since martial law was lifted in 1987, crowds have gathered annually to commemorate those who disappeared or were killed in 1947. Numerous books and articles have been written on the incident, and it has even been the subject of academic conferences as modern-day Taiwanese endeavour to find out exactly what happened in that year. More and more people have come out into the open to describe what they saw during the massacre. For the opposition, the incident has become a symbol of KMT oppression and the sacrifices made by the Taiwanese in their fight for freedom and dignity. Legislators of the opposition DPP pressed the government to apologize for the incident, to compensate the families of the victims, and to declare 28 February a day of national reconciliation. These efforts resulted in the erection of monuments in several major cities (including one in a public park in Taipei), while government officials began to take part in the memorial gatherings. The February 28 Incident, which had intimidated the Taiwanese into silence for more than thirty years, has now become a rallying point for the opposition in their attempt to push the government towards further democratization.

Although brutality on the scale of 1947 seems unlikely now that the public have acquired more peaceful channels to express their dissatisfaction and the government appears determined to move forward on the road to democracy, the division between Taiwanese and mainlanders has never lost its political relevance. On the contrary, the division demonstrates the very complexity of the island's politics. Since the mid-1970s, when Chiang Ching-kuo came to power, the KMT has tried to recruit the Taiwanese business and political élite into its decision-making circle. And because of the regime's deep penetration into grass-roots social organizations and the bureaucracy, the majority of the local élite has been co-opted into the ruling party. As a result, by the early 1990s Taiwanese constituted about 70 per cent of KMT membership. In addition to this, government control of the media and the education system effectively persuaded the people that the KMT was the legitimate ruling party. Consequently, although the DPP appears to be the obvious choice of the native population, it continues to find it difficult to win the support of all Taiwanese. As Chapter 4 will demonstrate, the key factor that divides the KMT and the DPP is ethnicity, but the difference is that the DPP is clearly a Taiwanese party while the KMT is a catch-all party.

The issue of provincial origin also plays an important role in KMT internal politics. The most prominent example of this is the way that the two main factions of the party, which showed the first signs of division in 1990, divide roughly along Taiwanese and mainlander lines. The mainstream faction, which represents the voice of the native population, argues that the trend of 'Taiwanization' initiated by Chiang Ching-kuo should be continued and the government should pursue a policy that sets the welfare of the people in Taiwan as its priority. The non-mainstream faction, consisting of mainlanders in the military, the KMT apparatus, veterans' groups, and academics, has progressively drifted away from the political forces that represent the Taiwanese population. As this faction was about to be sidelined by President Lee Teng-hui at the fourteenth party congress in 1993, some of its members formed the New Party, which gained support chiefly among mainlanders.

When Lee Teng-hui succeeded Chiang in January 1988, he had the strong support of most Taiwanese because he was the first of their number to reach the top of the political ladder. He has continued to enjoy a high level of popularity among Taiwanese, and this has enabled him to weather a few periods of political turbulence and to push through the reforms that have destroyed the power of the KMT old guard. Lee's being Taiwanese has also defused some of the DPP's most threatening charges, such as that the KMT is a 'mainlander' regime and Taiwan should pursue a policy of independence from mainland China.

In contrast to the politics of ethnicity, the society in general reflects a different picture. Differences between Taiwanese and mainlanders have been blurred by forty years of coexistence. The children of mainlanders and Taiwanese go to the same schools, receive the same education, and the men serve together in the military. The adoption of Mandarin as the official language and the only spoken language in the education system has greatly improved communication between the two groups. Competitive examinations ensure equal access to higher education, the cost of which is deliberately kept low. The mixing of the two groups in the education system also makes intermarriage between Taiwanese and mainlanders more common and acceptable to people in Taiwan today, and this has done a great deal to promote ethnic homogenization (Tsai, 1987: 6). Moreover, because of their long physical separation from China, many mainlanders, particularly the younger generation born in Taiwan, strongly identify themselves with the island (Tsai, 1987: 7).

The economic sphere is another matter. Most public and semi-public sector companies are headed by retired mainlander officials and military officers (or their children), given these jobs as rewards for loyal service to the regime. The most remarkable examples are China Airlines, which serves as a retirement home for air force generals, the three television networks, and the Chinese Petroleum Corporation. Although staff at middle management level and below are recruited on merit, these quasi-monopolies remain basically the main-landers' turf. The picture is quite different in the private sector, however. With a few notable exceptions, such as the mainlander-run Hualon group, private companies are gener-ally owned and operated by Taiwanese. Indeed, most second generation mainlanders feel excluded from the private sector due to their inability to speak Taiwanese dialect, and this has engendered a strong feeling of victimization among young mainlanders.

In short, ethnicity is still a problem in Taiwan, but it is not just a simple problem of one group dominating the other. As the polity becomes more democratic, the KMT will no longer penetrate so deeply into society and the old mechanisms of political control will gradually break down. As a consequence, those mainlanders whose power and privilege depend on authoritarian rule will lose out. The division between Taiwanese and mainlanders will continue to be politically significant, however, as long as mainlanders are trying to preserve their power in a political system that is increasingly dominated by Taiwanese.

Taiwan's China Tie

Taiwan's ties to China date from 1281 when the Mongol Yuan dynasty established an official presence in the Pescadores (Penghu), which lie between the mainland and Taiwan (Mancall et al., 1964: 43). Chinese migration to the island started in the sixteenth century, when hunger, poverty, and overcrowding in southeast China drove people in search of a more fertile and less crowded land. Meanwhile, the Western sea powers, including Portugal, Spain, and the Netherlands, had also landed on the island, attempting to make Taiwan their base for trad-ing with China. In 1642, the Dutch East India Company expelled the Spanish settlers and became the rulers of Taiwan (Davidson, 1903: 22). Dutch missionaries were already active on the island by this time. By 1635, 700 aborigines had been baptized by the Dutch, and in 1636, they established the first school on the island (Davidson, 1903: 25).

The Dutch did not rule for very long, however. Zheng Chenggong (known to the Dutch as Koxinga), a military commander loyal to the recently toppled Ming dynasty, expelled the Dutch settlers in 1661 and brought with him the largest group of Chinese migrants to the island (Davidson, 1903: 50–1). Taiwan eventually came under the rule of the new Qing dynasty and the Chinese population increased steadily up to the time when Taiwan was ceded to the Japanese in 1895. Taiwan thus became ethnically and culturally Chinese.

Although Taiwan was annexed by the Qing empire in 1683, the Manchus did not pay serious attention to the island, and the conflict between the Chinese migrants and the aborigines, and between Fukienese and Hakka, continued for two hundred years. It was not until 1887 that the imperial court realized the economic and political importance of Taiwan and made the first attempt to bring some order to the island. Taiwan was declared a province of China and Liu Mingchuan was appointed its first governor (Mancall et al., 1964: 47). Liu undertook major social and economic reforms, strengthening the island's military establishment, developing the transportation and communications systems, modernizing the education system, and carrying out an extensive land survey (Mancall et al., 1964: 48).

It is argued by some Taiwanese exiles in the United States and some opposition politicians that Taiwan has never been effectively ruled by China. They claim that the island was abandoned by China in order to make peace with Japan, and right after its 'return' in 1945 it was separated from China again because of the defeat of the Nationalist government (Yao, 1988: 8–9). Since Taiwan's connections with China are so tenuous, they argue, it should not be ruled by any Chinese regime, Nationalist or communist.

The legal status of Taiwan and whether it is a part of China is still under debate. Historical records show that the United States agreed to Chiang Kai-shek's proposal at the Cairo Conference of 1941 that Taiwan be returned to China after the war ended, and this agreement was confirmed in the 1945 Potsdam Declaration on the eve of the Japanese defeat (Copper, 1990: 25). The declaration did not specify, however, which Chinese government—Nationalist or communist—would have jurisdiction over the island. Chiang's forces did occupy Taiwan after the war ended, but his government subsequently lost control over the Chinese mainland. The logic of the Taiwan independence advocates is that the Nationalists, who took over Taiwan, lost jurisdiction over China and the communists never ruled Taiwan anyway, therefore Taiwan is legally not a part of China (Yao, 1988: 12). The Nationalists, for their part, argue that they have never abandoned their claim to sovereignty over the Chinese mainland; they were the sovereign government of China when Taiwan was returned to Chinese rule and therefore they should have jurisdiction over Taiwan.

The international status of Taiwan continues to be debated between the opposition DPP and the ruling KMT, and has become the most controversial issue in Taiwan's domestic politics. This debate will be discussed at length in Chapter 5.

Taiwan under Japanese Rule

In 1895, China entered into a disastrous war with Japan which ended with Taiwan and the Pescadores being ceded to the Japanese under the Treaty of Shimonoseki. After the treaty was signed, there was a short-lived attempt to offer military resistance to the occupying Japanese forces. As news of the treaty reached the island, Governor Tang Jingsong proclaimed Taiwan an independent republic and himself the first president, and vowed to return Taiwan to China's rule when the Japanese were expelled (Davidson, 1903: 278). When Tang was defeated in the north, Liu Yongfu, the military commander in southern Taiwan, reorganized the resistance and was elected by his followers as the new president (Davidson, 1903: 351). But the resistance was no match for Japan's naval power, and the fighting did not last long. Pockets of resistance were put down with heavy losses on the Taiwanese side (Kerr, 1974: 112). Within a few years, the Japanese had completely pacified Taiwan.

In spite of the brutality of colonial rule, the Japanese did bring to Taiwan law and order, which had been largely absent in the past. The Taiwanese were eventually granted a limited degree of local autonomy, and small-scale elections were held beginning in 1935. The elections were hardly fair, however, as Japanese far outnumbered Taiwanese in the provincial and district assemblies. There were Taiwanese majorities only at village level and in one municipal assembly (Ballantine, 1952: 28). Despite these drawbacks, the elections did give Taiwanese their first limited exposure to democratic processes, something which most other colonial populations were denied. Many members of the local Taiwanese élite were able to use the elections to attain a degree of political power, and they seemed to get into the habit of representational politics during the years of colonial rule. When the Nationalists retook Taiwan in 1945, many Taiwanese assembly members demanded continued autonomy or local self-rule, to the irritation of the Nanking government.

The most important achievement of Japanese rule, however, was a high degree of socioeconomic development together with the construction of a modern infrastructure and irrigation system. Anyone who has studied Taiwan's economic development in the 1970s and 1980s would certainly agree that the Japanese laid a very good foundation for the Nationalist government's economic development measures (Gold, 1986). Wanting to make Taiwan a military base for conquering Southeast Asia, the Japanese took special care to develop the island. Thomas Gold has summarized the achievements of the Japanese as follows:

[The Japanese] took the lead in creating what today would be called a good investment climate on the island: enforcing law and order; unifying weights, measures, and currency; guaranteeing private property rights; building a modern infrastructure; mobilizing natural resources; increasing agricultural productivity; making investment capital available; and developing human capital, including the provision of public education and employment for women. (Gold, 1986: 44–5)

By the time the Nationalists took over in 1945, Taiwan already had an island-wide road system, railroads, irrigation networks, electric power, public sanitation networks, and perhaps most important of all, public education facilities. The level of development was much higher than that of the Chinese mainland.

Unlike most other former colonies, Taiwan's agriculture was not entirely organized on a plantation system (Gold, 1986). The high degree of social and economic polarization engendered by large plantations employing huge numbers of impoverished labourers makes it difficult for former colonies to construct a self-sufficient agricultural system. At the time of the Nationalist takeover, the land was mostly in the hands of small and medium-sized landlords and individual farmers (Peng, 1973). The Japanese-built factories were scattered around the island, meaning that there was no proletariat as such. Neither were there any Taiwanese capitalists to speak of, as locally-owned businesses were small and few in number (Gold, 1986). In short, economic inequality among the Taiwanese was not pronounced at the time of the Nationalist takeover. This gave Taiwan a better start than most other Third World countries, many of which had to battle poverty and inequality from the beginning of their development.

The Chiang Kai-shek Era

Chiang Kai-shek ruled Taiwan from the time of the Nationalist government's retreat from mainland China in 1949 until his death in 1975. The government structure under Chiang was based on the Chinese constitution of 1946 which provided the island with strong government institutions able to maintain political order and implement economic policies. The nature of the regime could best be characterized as quasi-totalitarian or tight authoritarian, and individual political freedom was sacrificed in order to maintain the myth of recovering the Chinese mainland. Elections continued to be held at local level, however, and these prepared the Taiwanese for a more democratic future.

General Characteristics of the Regime

Despite the elections, the Chiang Kai-shek era has been described by Cheng (1989) as 'quasi-Leninist' and by Winckler (1984) as 'hard authoritarian' for its comprehensive control over society, the structure of the ruling party, and relations between party and state. The degree of control assumed by Chiang was close to that found in a Leninist state, but it differed in two major areas: the existence of private ownership and market exchange and the institutionalization of local elections (Cheng, 1989: 477–8). Under the rigid political system, Chiang was the head of state, commander-in-chief of the military, chairman of the ruling party, and the final arbiter of government policies. Martial law was in force, which gave the Taiwan Garrison Command almost unlimited freedom to police the state and society, while the Kuomintang had penetrated deeply into the most distant corners of the island and firmly established near-absolute control.

Under Chiang's rule, people were never allowed to criticize the Generalissimo himself, his family, his policies, or his government. Punishment was severe for those who dared to disagree with him. Moreover, a network of secret service agents kept a lookout for signs of dissent. There were spies in high schools and colleges, and military officers were permanently stationed in schools to oversee and discipline the students. There were also personnel in charge of security in every government office, school, and large factory to ensure the loyalty of civil servants, teachers, and factory workers.

As in any Leninist state, there was a highly visible personality cult. Chiang was depicted as a godlike figure fighting a holy war against communism on behalf of the Chinese nation. His statues were to be seen everywhere on the island, reminding people that he was firmly in power and no one else could ever surpass his position in the state. Recovering the Chinese mainland from communist rule was the highest national goal, and in order to achieve it a large military establishment was maintained with all adult males required to complete a period of military service. The KMT was organized as a revolutionary-democratic party and the politica freedoms and civil liberties guaranteed by the constitution were restricted. Some liberal-oriented intellectuals, chiefly mainlanders centered around the *Free China Fortnightly* (*Ziyou zhongguo*), did attempt to institute an opposition party in cooperation with some members of the local élite in 1960, but the attempt ended in mass arrests by the regime. This crackdown set the limit of government toleration of political dissent and effectively curbed any further opposition activity. No articulation of opposition views was permitted at any time under Chiang's rule.

The Constitution

The constitution of Taiwan was drafted and adopted in mainland China during the civil war between the Nationalists and
the communists. In 1945, the Japanese invaders surrendered
to the Allies and the war between Japan and China ended, but
the Chinese civil war, which had been temporarily halted during the Japanese invasion, resumed and intensified. By 1946
it was apparent that the Nationalists were losing the war, so
Chiang Kai-shek, in order to regain legitimacy in the eyes of
the Chinese people, called a meeting of national reconciliation in Nanking. After a series of negotiations, this meeting
endorsed a draft constitution based on Sun Yat-sen's *minquan*
principle. This document was later enforced in Taiwan by
Chiang as a symbol of his regime's claim to China.

Under this Constitution, laws were to be made by the
Legislative Yuan, the Control Yuan was to oversee the ethics
of government officials and civil servants, and the National
Assembly, the highest organ of popular sovereignty, was em-
powered to elect and recall the president and vice-president,
initiate and veto legislation, and amend the Constitution.
Members of both the National Assembly and the Legislative
Yuan were to be directly elected by the people, and members
of the Control Yuan were to be elected by city councils and
provincial assemblies. The above three branches of the central government together form the three branches of the
Parliament (see Figure 2.1).

The heads of the Judicial Yuan and Examination Yuan were
to be nominated by the president and approved by the Control
Yuan.[2] The head of the Executive Yuan, or premier, was to
be nominated by the president and approved by the Legislative
Yuan. Ministers and other high executive officials were also
required to report to the Legislative Yuan on important government policies and to be subject to interpolation by Legislative
Yuan deputies. This design is somewhat similar to the British
parliamentary system, except that the legislature has a specific term and the legislative and executive branches are
separated. Furthermore, a checks-and-balance mechanism exists
between the two major branches of the government. In addition, because the president is also the commander-in-chief,
he has a tremendous degree of autonomous power free from
legislative supervision. In short, the system's institutional
appearance is parliamentary, in the sense that the executive

Figure 2.1 Structure of the Government of the Republic of China, 1947 Constitution

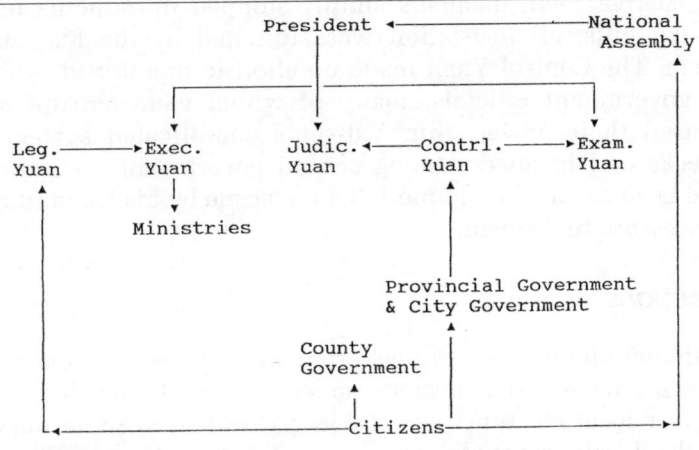

—→ represents election, appointment, or confirmation of appointment.

is responsible to the legislature, but it is presidential in the sense that the president has real political power and a system of checks and balances exists between the executive branch and the legislature.

In practice, the parliament did not, and apparently could not, exercise much power, and the mechanism of checks and balances stopped functioning in the way it was designed to. Because of the domination of a single party over the state and the domination of one man over the party, Taiwan inclined strongly towards a presidential system. Throughout the KMT's rule, with the exception of the brief interlude of Yen Chia-kan's presidency after the death of Chiang Kai-shek in 1975, the president has always been the chairman of the ruling party. Furthermore, because the KMT always had a majority in the parliament, the members of that body were easily persuaded to give up their constitutionally designated autonomy and become rubber stamps of the executive. The National Assembly, designed by Sun Yat-sen to represent the people and to directly control the government, was convened only one week per year for ceremonial purposes, and once every six years it went through the motions of electing the president and vice-president. With an original membership of more than three thousand (more than one thousand of whom retreated to

Taiwan with the Nationalists), the National Assembly was too cumbersome to make any important decisions on its own. Legislative Yuan members simply stopped introducing individual bills; all legislation were initiated by the Executive Yuan. The Control Yuan made no effort to monitor the ethics of government officials, many of whom were corrupt and abused their power. Sun Yat-sen's complicated system of checks and balances among central government institutions had been reduced to no more than a facade by his Kuomintang successors in Taiwan.

Elections

Although the politics and government of the island could not be characterized as democratic under President Chiang Kai-shek, regular local elections (see Table 2.1) did have some impact on the development of democracy in Taiwan. This gave Taiwan a head start compared to many other Third World countries.

In spite of the frequent charges by opposition politicians that the KMT systematically conducted electoral fraud and vote-rigging prior to the lifting of martial law, the local elections appear to have had a moderately positive impact on the development of democracy. They did at least enable people in Taiwan to understand and appreciate the value of participation in a modern society. The elections also gave members of the local élite a chance to compete with one another at the ballot box. One advantage for the government, however, was that the elections probably diverted people's attention away from national affairs to issues that would not immediately challenge or jeopardize the legitimacy of the government. Moreover, the emergence of local factions in these campaigns gave the KMT even more leverage to manipulate local affairs, as the factions had to demonstrate their loyalty to the KMT in order to have their candidates nominated and thus retain their power. KMT candidates supported by local factions were usually able to sweep aside individual members of the opposition.

National elections, in contrast, had a much less positive impact. The first presidential election in the Republic of China was conducted in Nanjing in 1948 by the newly elected National Assembly. Chiang Kai-shek easily defeated the only other candidate for the presidency, but competition was intense among the six contenders in the vice-presidential race and it was

Table 2.1 Local Elections in Taiwan, 1950–1994

Year	Provincial Assembly	Taipei[a] Council	Kaohsiung[a] Council	County[b] Councils	County[b] Magistrates
1950				X	X
1951	X				
1952				X	
1954	X			X	X
1955				X	
1957	X				X
1958				X	
1960	X				X
1961				X	
1963	X				
1964				X	X
1968	X			X	X
1969		X			
1972	X				X
1973		X		X	
1977	X	X		X	X
1981	X	X	X		X
1982				X	
1985	X	X	X	X	X
1989	X	X	X	X	X
1993					X
1994	X	X	X	X	

Notes: [a] Elections in Taipei and Kaohsiung were considered county-level elections until the two cities were brought under the direct jurisdiction of the Executive Yuan.
[b] County-level elections include counties and county-level cities.

Sources: Compiled from ROC Central Election Commission, Summary of Elections in the Republic of China (Taipei: CEC, 1984), Vol. 2, and newspaper reports published in Taiwan after 1984.

only after the fourth ballot that Li Tsung-jen was declared the winner (CEC, 1984, V1: 3–8). The second presidential election took place in Taipei in 1954. There were two candidates for the presidency and three for the vice-presidency. After two ballots for each, the winners were Chiang Kai-shek and Chen Cheng.[3]

According to the 1947 constitution, the president can serve
for only two six-year terms. Chiang's second term expired in
1960, but this obstacle was circumvented by amending the
Emergency Decree to allow him an unlimited number of
terms. (CEC, 1984, V1: 22). In the 1960 presidential and vice-
presidential races and thereafter, there was only one candidate
for each post. Thus presidential elections became six-yearly
rituals and all democratic significance was lost. Chiang was
president of the Republic of China up to his death in 1975.
In the 1978 election, Chiang Ching-kuo, Chiang Kai-shek's son,
was elected president and Shieh Tung-min vice-president. The
younger Chiang was re-elected in 1984, and his vice-president
was Lee Teng-hui, the current president of the ROC.

No parliamentary elections were held in Taiwan before 1969.
After the government transferred to Taiwan, members of the
National Assembly, the Legislative Yuan, and the Control Yuan
were frozen in office, with the seats of those members who
did not cross the Strait, or those who died in the meantime,
filled, wherever possible, by runners-up in the original 1947–8
elections. In the National Assembly alone, if the first runner-
up was not available, the candidate who came in third could
be substituted (CEC, 1984, V1: 165).

This practice of 'filling-in' (dibu) the parliament with
runners-up was generally ridiculed by Taiwanese. It was pointed out that the runners-up had, by definition, been rejected
by the electorate in the 1947–8 elections, they were often
nobodies, swept from obscure positions overnight, and what
is more, the system guaranteed the continued mainlander domination of the parliament. Taiwanese also complained that
while they had to compete with each other for the few supplementary seats available in the parliament after 1969, some
mainlander members had bypassed the electoral process completely. The issue was seriously debated in the Legislative
Yuan after martial law was terminated, and the relevant clause
of the National Assembly Election Law was finally suspended on 11 March 1988 (The Journalist, 58: 22). Some opposition leaders further argued that the runners-up should be
expelled from the parliament, but their arguments were to no
avail.

In 1969, the government responded to the rapid increase in
Taiwan's population by holding the first of a series of sup-
plementary elections. Eight new members were elected to the
National Assembly and eleven to the Legislative Yuan (CEC,

1984, V1: 206, 598–601). Just like their mainlander colleagues, members elected in these first supplementary elections were not required to run for re-election. Although the 1969 election was on a very small scale, it was significant in that for the first time people in Taiwan were able to participate in national-level politics, and the candidates were able to discuss issues that concerned the entire nation.

The supplementary elections of 1972 were held on a larger scale, and 119 candidates were elected to the three branches of the parliament. Of these, only 38 National Assembly members and 28 members of the Legislative Yuan were elected directly (CEC, 1984, V1: 246–8), the rest were elected by professional organizations and overseas Chinese. Of course, these functional constituencies were controlled by the KMT to safeguard its absolute majority in the parliament. Supplementary members of the Control Yuan were elected indirectly by the city councils and the Provincial Assembly. These supplementary elections were institutionalized and became the most important electoral contests in Taiwan. Opposition politicians were able to enter the national political arena as non-partisans and challenge government policies. Credit for expanding the scale of supplementary elections must be given to Chiang Ching-kuo rather than his father, however, as the younger Chiang, who became premier in 1972, initiated the gradual liberalization while his father was ill.

With the exception of local elections, there was virtually no political freedom nor were there any truly democratic institutions in Taiwan under Chiang Kai-shek. It was to prove very difficult for Taiwan to disentangle itself from the comprehensive legal and political web of authoritarianism and transform itself into democracy in the second half of the 1980s.

Summary Conclusion

History frequently gives important clues to modern events. From the above account, it is clear that the democratization of Taiwan is the result of a continuous effort by the Chinese to find a political system that is both democratic and adapted to conditions in China. Sun Yat-sen devised an ambitious and complicated democratic system of government that incorporated both popular representation and checks and balances.

However, the Nationalist revolution and all hope of establishing a modern republic was shattered by the activities of the warlords, the Japanese invasion, and the civil war with the communists. After the defeated Nationalist government retreated to Taiwan, distrust between Taiwanese and mainlanders and the oppressive nature of the regime made it very difficult for people in Taiwan to call openly for democracy during the first few decades of KMT rule.

In the subsequent transition to democracy, Taiwan's ambiguous relations with China and the debate over whether Taiwan is really a part of China, became one of the most important factors. On the one hand, the Nationalists presented themselves as the legitimate government of the whole of China which they were preparing to retake from the communists by force. This cause was used to legitimize the prolongation of KMT rule and the unchecked power of Chiang Kai-shek and to justify the need for martial law and the restriction of civil liberties. Relations between China and Taiwan also entered the debate on Taiwan's political transition, as people questioned whether Taiwan should unite with China and if so, how the new system should incorporate the national goal of unification without compromising the substance of democracy.

However, by continuing to uphold at least the ideal of Sun Yat-sen's vision of democracy and promoting rapid economic development in Taiwan, the KMT in some sense worked against its own long-term interests. As economic and social conditions improved, many people, particularly intellectuals and members of the growing middle class, began to press the government for more freedom and full implementation of the constitution. The following chapter will describe why the KMT began loosening and losing its tight grip on the state, how the economy and society were transformed by the government's modernization policy, and why people began demanding political rights and freedoms.

Chapter 3 ∎

Forces for Change: Social, Political, and Economic

THE KMT's Leninist-style grip on Taiwan began to show signs of relaxation in the early 1970s when Chiang Kai-shek's son, Chiang Ching-kuo, was elevated to political prominence. Freedom of expression gradually increased and the process of controlled liberalization and democratization culminated in the abolition of martial law in 1987. However, the political transition on the island was not determined by a single factor alone. Chiang Ching-kuo's decision to liberalize may have been the key to the relatively peaceful nature of the transition, but social and economic changes, particularly social pluralization and the increase in the standard of living that resulted from Taiwan's development in the 1970s and 1980s, were too important for the regime to ignore; they could no longer resort to the old mechanisms of control and repression. In other words, it is necessary to look at both the decisions taken by the regime and the social forces working for change in order to understand Taiwan's political development. Such an examination may provide some insights into the phenomenon of democratization, though it is not the intention in this chapter to construct any grand theory.

Chiang Ching-kuo: The Turning Point

The Chiang Ching-kuo era in Taiwan politics began in 1972 when the younger Chiang became premier. Although his father was still considered the paramount leader of the country, Ching-kuo was undoubtedly making all the important policy decisions on his own, for his father was ill most of the time during his final years. Ching-kuo's rule in Taiwan can be divided into three periods. The first (1972–78) was characterized by a loosening of the quasi-Leninist system, the second (1979–85) was a period of a soft authoritarianism, and in the final period (1986–88), Taiwan appeared to have embarked on the democratization process.

One distinction between these three periods may be seen in the way the regime dealt with the opposition. In the years from 1972 to 1978, almost no dissenting voice could be heard and only a few oppositionists could enter politics as individual non-partisans. After 1979, some opposition journals were permitted and the non-partisans were allowed to form an island-wide coalition under the close scrutiny of the regime. Beginning in 1986, the opposition was allowed to criticize freely and could actually form a political party to challenge the government without fear of a crackdown.

The First Period (1972–1978)

Although Chiang Ching-kuo assumed power as a continuation of Chiang Kai-shek's rule, there are important distinctions between the two. Unlike his father, who suppressed all forms of opposition, Ching-kuo was willing to gradually liberalize the system. Soon after the younger Chiang became premier in 1972, a group of professors and students at National Taiwan University founded a journal, *The University* (*Daxue*), to advocate political reform and liberalization, a move actually tolerated and even quietly supported by the KMT.[1] In 1975, Chang Chun-hong, a KMT cadre who was dissatisfied at his failure to be nominated by the party for a Provincial Assembly seat, withdrew from the party and founded *Taiwan Tribune* (*Taiwan zhenglun*), the first journal to discuss sensitive political issues since *Free China Fortnightly* was banned by Chiang Kai-shek in 1960.

More significantly, the supplementary elections began to be held regularly and on a larger scale after 1972. More and more non-partisan politicians were able to enter parliament and challenge government policies. A few native Taiwanese, such as Shieh Tung-min, Lin Chin-seng, and Lin Yang-kang, were even invited into the ruling circle by Chiang Ching-kuo as part of his 'Taiwanization' policy aimed at winning over the local élite.

Nevertheless, it would be wrong to suggest that there was

a movement towards democracy in Taiwan during the early years of Chiang rule, for much evidence suggests otherwise. At the structural level, the KMT was still the overwhelming political force in Taiwan and no other force was even allowed to take shape to challenge it. The supplementary elections did not change the fact that the majority of the parliamentarians were frozen in office. The parliament functioned, very much like the Supreme Soviet in Moscow and the National People's Congress in Beijing, merely as a rubber stamp. There were too few opposition members for them to have any influence. While Ching-kuo appeared to be more open-minded than his father, there was no doubt that he too was a strongman, and all important policy decisions, administrative or economic, were approved by him personally. At a more basic level, freedom of speech was still curbed by the regime, and political persecution was still common.[2]

In the run-up to the supplementary election of 1978, opposition politicians for the first time agreed to run a joint campaign. As a consequence, it was predicted by many political observers that they would make a strong showing. However, the sudden announcement by the Carter Administration that the United States would switch diplomatic recognition to the People's Republic of China (PRC) prompted the government to postpone the election (CEC, 1984, VI: 370). Nonetheless, the opposition cause continued to gather steam. In 1979, the opposition formed itself into an integrated political force around the magazine *Formosa* (*Meilidao*). *Formosa* drew together all the important opposition leaders in Taiwan, and the term *dangwai* ('outside the party', i.e., non-KMT) became a synonym for an opposition party. The magazine was the propaganda machine for the Dangwai, while its local distribution centers served as 'party' branches and its subscribers were treated as potential members.

The rapidly growing opposition adopted a strategy of mass protests and seemed to be on a collision course with the KMT. After a mass rally in Kaohsiung on 10 December 1979 developed into a riot, the government clamped down.[3] The subsequent mass arrests encompassed almost all major Dangwai leaders and their most active supporters. After the trial in a military court in early 1980, forty-seven non-partisan opposition politicians were jailed. Shih Ming-teh, the general manager of the magazine, received a life sentence for multiple sedition offences,[4] and other founders of the journal received twelve- to fourteen-year sentences, all for sedition. The magazine was shut down permanently after publishing only four issues.[5] Chiang Ching-kuo had made it clear to the opposition that there was a limit to the regime's toleration of its activities, and the formation of a new political party was beyond that limit.

The Second Period (1979–1985)

Beginning in 1979, a greater degree of political freedom was gradually permitted in Taiwan. The opposition was dealt a severe blow by the ruling KMT after the Kaohsiung incident of December 1979, but it was not totally wiped out. The KMT continued to permit the publication of some opposition journals, and a moderate force led by Kang Ning-hsiang and his magazine *The Eighties* (*Bashi niandai*) continued to serve as a mouthpiece of dissent. When elections resumed in 1980, members of this moderate opposition and close relatives of the jailed political leaders were able to win seats in the parliament. In subsequent years, the Dangwai seemed to be testing and pushing the KMT to the limit of its tolerance. The opposition was allowed to challenge the KMT in the parliament, but the ruling party was adamant that it would never relinquish power. There was no sign that the opposition would be allowed to form a true political party. Although the regime was suspected of being linked to three political murder cases during this period,[6] in general terms Taiwan did appear to be enjoying a more liberal political climate in the second period of the Chiang Ching-kuo era.

The government contends that the process of democratization started in 1984–85, on the grounds that it did not crack down during that period on opposition groups such as the Dangwai Association for Campaign Support, the Dangwai Writers and Editors Association, and the Dangwai Public Policy Research Association. But foreign press reports of systematic violations of free speech (AWSJ, 7 January 1985: 6) and political persecutions (AWSJ, 26 July 1985: 6) as late as July 1985 seem to argue against that dating. It could well be that the KMT leadership itself had yet to reach a consensus on the issue of liberalization.

The Third Period (1986–1988)

Democratization in Taiwan was formally initiated in March 1986 during a meeting of the Kuomintang's Third Central Committee. Chiang outlined six areas of political reform and revealed his intention to open up the system. The event certainly demonstrated Chiang's determination to put Taiwan on a democratic course. Many people were still sceptical about the prospects for democratization, however, because despite Chiang's announcement no major reforms materialized. Strong resistance seemed to exist within the KMT against the move toward democratization, and the government-controlled newspapers were giving conflicting signals to the public. The ambiguity was cleared by Chiang on 7 October 1986 in an interview with Katharine Graham, publisher of the *Washington Post*, during which the aging president disclosed for the first time in public his determination to lift martial law and possibly permit the formation of new political parties. The interview eased the fears of the opposition, who had formed the DPP on 28 September in defiance of the government's long-standing ban on forming political parties.

Included in Chiang's package was reform of the parliament. His determination to carry this out was revealed by a government official close to the president personally. The official said that the reason parliamentary reform made slow progress at first was that it was being supervised by the former president, Yen Chia-kan, who was confined to a hospital bed most of the time. Chiang regretted the delay, the official said, and worrying about that aspect of reform had probably hastened his death in January 1988 (personal interview, 10 September 1988).

An important aspect of the liberalization was the way the authorities turned a blind eye to the DPP, which was technically illegal. Many conservative KMT officials argued that the opposition should be punished severely for defying the law, and the *Central Daily News*, the organ of the KMT, carried articles and editorials sharply criticizing the opposition and urging the government to take swift action. Surprisingly, the government took no action at all. It was widely reported that it was Chiang who single-handedly held back the surging tide of conservatism inside the KMT (*Central Daily News*, 31 January 1988). This could well be the case, as Chiang had mentioned the possibility of allowing opposition parties in his *Washington Post* interview.

Perhaps the contrast between the final period of the Chiang Ching-kuo era and the earlier periods is most marked in the behaviour of the Legislative Yuan after 1986. The thirteen DPP legislators made a formidable effort to challenge KMT policies, criticizing the government's stance on various issues. They kept the KMT up to the mark on parliamentary reform, and bargained and negotiated with their KMT colleagues for the passage of several pieces of legislation, including the National Security Law, amendment to the Election Law, the Civil Organizations Law, and the Public Demonstration and March Law.

Very frequently the opposition legislators would use their right to make statements as an opportunity to humiliate the government over its failure to hold general elections and for retaining the pre-1949 legislators. They continuously pressed for the resignation of these senior legislators. Sometimes they would go so far as to start fist fights with KMT legislators just to put their point across. Their strong attacks forced the government to take a softer stance on some legislation. In particular, Ju Gau-jeng, at that time a DPP legislator,[7] caused the KMT a great deal of trouble. He was known for jumping onto the podium, tearing up documents, arguing with KMT officials, and humiliating elderly members. Ju's antics in the Yuan were a media sensation, and they won him the nickname 'Taiwan's number one battleship' (*Taiwan diyihao zhanjian*). In this period, the Legislative Yuan, despite the fact that the majority of its members had never stood for election in Taiwan, began to exhibit signs of multi-party competition. The Yuan itself became the scene of a fierce struggle, as well as negotiations and compromises, between the ruling party and the opposition. In the 1950s and 1960s it seems, no one would have predicted that the legislature would take on this appearance in the final years of Chiang Ching-kuo's rule.

Chiang also attempted to deal with the sensitive issue of Taiwanese–mainlander conflict. Mainlanders were undeniably the dominant actors in Taiwan politics, but the government had always refused to admit that the Taiwanese majority was oppressed by the mainlander minority. On 10 May 1987, Chiang revealed to a group of KMT legislators that forty years previously his father had asked Chen Cheng, then the governor of Taiwan, to promote outstanding young Taiwanese to high positions (*The Journalist*, 21: 4). On 15 July, Chiang told a group of Taiwanese friends that as he had lived in Taiwan for forty years, he should be counted as a Taiwanese (*The Journalist*, 21: 4). Chiang was already noted for bringing more Taiwanese into the leadership, but these remarks demonstrate his willingness to confront the issue, and they encouraged Taiwanese to press for more equality in politics.

It is quite clear that during the third period of Chiang's rule, Taiwan was set on a firm course of political transition. Chiang's decision not to use repressive tactics against the opposition ensured that this transition process would be peaceful. However, one important question deserves attention: Why did Chiang decide to embark on reform rather than continue repression?

Events Prior to Political Reform

In order to answer the question of why Chiang started the process of reform in 1986, it is important to understand the significance of certain events that took place just prior to the decision to reform and discover what went wrong with the regime and Taiwan society during that period. In the first half of the 1980s, there were three incidents that might have provoked Chiang's concern over the political system: the murder of Henry Liu, the Tenth Credit Cooperative loan scandal, and certain spontaneous protest activities in central Taiwan. These events raised internal and external concerns over the regime's ability to deal with politics, the economy, and society.

Murder of Henry Liu

Henry Liu, a Chinese immigrant to the United States and the author of a biography of Chiang Ching-kuo that was banned in Taiwan, was found shot to death at his California home on 15 October 1984. The subsequent murder investigation implicated some of Taiwan's top military intelligence officers, a powerful organized crime ring, the Bamboo Union, and possibly Chiang's second son, Hsiao-wu (AWSJ, 27 March 1985: 10).

The government was publicly humiliated by news reports linking Taiwan's military intelligence with an organized crime ring. At first, it tried to distance itself from the murder and the Bamboo Union, but when all fingers were pointed at the KMT, the government quickly indicted two top military intelligence officers (AWSJ, 27 March 1985: 3) and handed out stiff penalties to the gangsters who allegedly carried out the crime (AWSJ, 10 April 1985: 3). Meanwhile, Chiang Hsiao-wu was packed off to Singapore as Taiwan's representative, away from the center of one of the worst political storms in Taiwan's history.[8]

The US government was understandably very annoyed over the KMT's murder of an American citizen in its own territory, and threatened to cut off arms sales to Taiwan, which was almost exclusively dependent on US armaments for its national defence. The regime was even more humiliated when a *Sixty Minutes* program portrayed the KMT in a very negative light and questioned the government spokesman, Chang King-yuh, on the Taipei government's role in the murder. It is still not clear whether Chiang Ching-kuo or his son were involved in the decision to assassinate Liu, but what is certain is that the case damaged the image of a regime which was trying very hard to present itself as a miracle of economic development that was willing to open up politically.

Loan scandal

Early in 1985, Taiwan's financial markets were rocked by revelations of a loan scandal involving the Tenth Credit Cooperative, owned and operated by Tsai Chen-chou, a KMT legislator and a boss of the prominent Cathay Group (*Guotai jituan*). On 14 February, the co-operative was ordered by the government to suspend its lending operations at a time when it was on the verge of bankruptcy through overlending. The suspension started a run on the co-operative and one of its sister companies, Cathay Investment & Trust Co., and sparked off the worst financial crisis and the worst non-political demonstrations Taiwan had ever experienced (AWSJ, 15 February 1985: 3).

Tsai was subsequently charged with fraud and other criminal activities (AWSJ, 15 May 1985: 9). The scandal also brought down two successive economics ministers, Hsu Li-teh and Lu Jen-kang, who were forced to resign for not foreseeing and stopping the criminal activities of the Tenth Credit Cooperative (AWSJ, 12 March 1985: 3). The financial crisis also prompted the government to take over the cooperative and Cathay

Investment & Trust Co., but these moves were not enough to prevent damage to the government's credibility.[9]

Protest activities

Despite martial law, there was an outburst of demonstrations in central Taiwan during this period over an environmental issue. As described by Hsu Mu-yuan, an activist who later became a Taipei City councilman, the government was unable to stop local groups from organizing mass campaigns to stop the US chemical company du Pont from building a plant in Lukang, halfway down the west coast of Taiwan. Among other activities, the people of Lukang invited scholars to deliver lectures on pollution and held demonstrations against the government's decision to license the du Pont plant (personal interview, 7 September 1988).

Activities in Lukang were widely supported by people elsewhere in Taiwan and for the first time an environmental issue got on to the national agenda (*The Journalist*, 2: 35). People began to recognize that two decades of industrialization without proper regard for the treatment of toxic waste had turned Taiwan into an area of heavy pollution (Severinghaus, 1989). The mass campaign in Lukang was used as a model by people in other cities seeking to protect their own interests, and demonstrations motivated by nonpolitical causes spread throughout the island.

Three Explanations

That Chiang's decision to reform was closely related to the above events is testified to by people close to him at that time. One of these is James Sung, now the governor of Taiwan and one of those who had direct access to the late president. He has spoken about the president's reform decision in an interview with *The Journalist*:

> I clearly remember that Mr. [Chiang] Ching-kuo took the decision to make the greatest breakthrough in politics in late-March 1986, after the Third Plenum of the KMT's Twelfth Central Committee. He seemed anxious. In a conversation with me, he said he felt that many problems remained to be solved, even though the Third Plenum was considered a success. Shortly afterwards, he gave instructions to [KMT] Secretary-General

Mah Soo-lay to start doing research right away on granting permission to form opposition parties, lifting martial law, reorganizing the provincial government, and reforming the parliament, so that they could be realized soon. (The Journalist, 95: 12)

In another interview, Ma Ying-jeou, now the minister of justice but at that time Chiang's English interpreter, confirmed that Chiang had appointed a twelve-person task force to study the six reform issues after the Third Plenum of the Twelfth Central Committee (*The Journalist*, 95: 14). According to these two interviews, the beginning of the transition process can be pinpointed at March 1986. Ma, in the same interview, directly supported the proposition that Chiang's decision to reform the system was related to the events discussed above. Ma said:

> In 1984, the case of Henry Liu's murder unfolded. Later, several large scandals, such as the Tenth Credit Cooperative scandal, were also disclosed. These cases hurt our international image to a great extent. This was why we decided in November 1985 to hold the Third Plenum when things calmed down a little. . . . The cases I just mentioned had a catalytic effect. They made him consider that all of these events needed to be discussed. (*The Journalist*, 95: 15)

This quotation strongly suggests that the direct cause of Chiang's decision was the social and political problems created by the murder of Henry Liu and the Tenth Credit Cooperative scandal, as well as other social problems.

Because Chiang's decision to reform the polity is so important to the entire process of democratization, the question why he opted for reform rather than repression needs to be answered. There are three possible explanations.

Personality

The first explanation is that Chiang was basically a benevolent ruler who wanted to do some good for his people and be remembered well by history. The basis for this explanation is that Chiang was sensitive to public criticism of the regime and the way he would be judged by historians after his death. The above two scandals and the social unrest had apparently shaken people's confidence in the government and public criticism of government officials had become relentless. Critics were pointing to the authoritarian structure of the state as the underlying cause of the chaos.

In contrast to the government's emphasis on both cleanliness and ability (*lian* and *neng*) in running public affairs, it was suddenly made public that the KMT government was closely linked with organized crime and was associated with one of Taiwan's wealthiest and most unscrupulous tycoons. Chiang, who always took criticism seriously, did not try to suppress the resulting outcry; instead, he adopted a policy of liberalization. Many people, including Sung and Ma, in the above interviews, express a belief that Chiang was very concerned about the welfare of the people of Taiwan, and that he undertook reform for their sake and to make the system more acceptable to them.

The ailing master of the regime

The second explanation is that Chiang sensed that he was in imminent danger of losing control over the authoritarian political system. This explanation is based on the assumption that Chiang wanted to maintain the KMT's dominant position over state and society in the long run. As Chiang's health weakened, some sections of the authoritarian bureaucracy, without a strong opposition to check their power, were beginning to act independently. In particular, the intelligence apparatus, which had been headed by Chiang earlier in his career, had become an independent bureaucracy running its own activities. The parliament, packed with representatives elected on the mainland whose average age was eighty-two in 1986, could not be expected to maintain the legitimacy of the KMT much longer. Chiang was the only person in the system powerful enough to push through the reforms and force the senior parliamentarians into retirement without an uproar from the conservatives. Chiang knew that his health would not get any better, and the only way to cope with the problem before it was too late was to institutionalize a democratic system that would remedy many of the regime's problems, including those of legitimacy and the need for a mechanism of checks and balances to oversee the operations of the bureaucracy.

Damage control

The third explanation, and perhaps the most persuasive one, is that Chiang saw reform as a way of limiting the extent of the damage the two scandals had inflicted upon the regime, both internally and externally. Externally, Washington's threat to cut off arms sales to Taiwan because of the Henry Liu murder was extremely serious as the Nationalist government relied entirely on the United States for sophisticated weapons to defend itself against China and for diplomatic support in the international community (AWSJ, 22 April 1985: 3). In these circumstances, Chiang saw political reform as a way of demonstrating to the Americans that the KMT did not have to resort to violence to deal with political opponents. Internally, with electoral politics becoming more and more competitive for the KMT, Chiang had to take people's minds off the scandals in time for the 1986 national-level elections. One basis of the government's legitimacy, the claim to represent the whole of China, had been seriously challenged, and it could not afford to lose its other props—US backing, continued electoral success, and outstanding economic performance.

It is impossible to say which of these three explanations of Chiang's decision is most plausible, since the president himself did not leave any statement or speech on the subject. Perhaps his motives were a combination of all three. But without further evidence (which seems unlikely to be forthcoming), one can only speculate on the reasons behind Chiang's decision to liberalize the polity.

The Lee Teng-hui Presidency

After Chiang Ching-kuo's sudden death in January 1988, his vice-president, Lee Teng-hui, succeeded to the presidency and the KMT chairmanship. The peaceful nature of the succession boded well for a continued transition to democracy. Lee brought more Taiwanese and Western-educated technocrats into the cabinet and the Central Standing Committee of the KMT in an effort to broaden the basis of support for the regime. For the first time in Taiwan's history, the number of Taiwanese in the cabinet exceeded that of mainlanders. President Lee, a native Taiwanese and Western-educated scholar himself, has shown no signs of wishing to retreat to the KMT's authoritarian past.

After Lee became president, measures were taken to carry

on the reform process initiated by the younger Chiang. A National Affairs Conference, the participants of which represented all important sectors and groups in society, was held in mid-1990 to discuss the most important political issues that faced the nation. The senior deputies in all three chambers of the parliament were forced into retirement at the end of 1991. The emergency decree was abolished, and the constitution was revised twice to bring it into line with Taiwan's political reality. Better relations were also achieved between the government and the opposition, ensuring that the DPP would not use violent means to overthrow the government and the existing political order. In general, Taiwan appeared to be moving steadily towards the goal of democracy under Lee's leadership. Indeed, it seemed that barring an attack from the mainland or a military coup against the civilian government—and these possibilities cannot be totally ruled out—Taiwan would be able to achieve a stable democratic system once the National Assembly had resolved the problem of how the president should be elected.

However, since Lee was not well connected to the old KMT establishment, he experienced difficulties in implementing the reform programs. Opposition came from within the KMT inner circle itself, particularly from those who had a vested interest in maintaining the party's monopolistic control over the state and society. When Lee was selecting his running mate in the 1990 presidential election, a group of conservatives was formed to challenge his position both in the government and the party. Lacking close followers in the establishment, Lee was forced to choose Hau Pei-tsun, an army general and former chief-of-staff, as premier in order to defuse the political crisis. One legacy of the crisis was the division of the KMT into a Mainstream faction (*Zhuliupai*), represented by Lee, and a Non-mainstream faction (*Fei zhuliupai*), represented by Hau. These two factions have continued to differ with each other over major political issues, and the KMT is threatened with a split.

Since Chiang Ching-kuo initiated the political liberalization process in 1986, Taiwan has come a long way. The achievement is all the more remarkable when compared to the mass protests and violence seen in other Asian countries such as South Korea and the Philippines, and the lack of political change in China or even Singapore. The major task for political scientists, then, is to look for the factors that are driving the political reform. Having discussed the personality factors above, it is time for a close examination of the economic and social basis of democracy in Taiwan.

Economic Growth and Income Distribution

Many political scientists have argued that socioeconomic development is the most important precondition of political democratization (Huntington, 1984; Lipset, 1959; Cutright, 1963; McCrone and Cnudde, 1967; Lerner, 1964; Deutsch, 1961). Briefly, these scholars assert that when socioeconomic conditions improve in a country, its citizens are more likely to make demands on the government for more political rights, thus making the political system more democratic. According to this theory, the most important basis for democratic political development is the advancement of the economy and the modernization of society. In order to find out whether Taiwan has the economic and social basis necessary for democratic political development, the first and most important aspect of Taiwanese society that needs to be examined is its economic development.

Economic Growth

According to the general indicators of economic growth, Taiwan has certainly achieved enormous success compared to its own past performance and that of other Third World countries (see Tables 3.1 and 3.2). Per capita GNP reached US$10,566 in 1993, and the average economic growth rate for the years 1952 to 1991 was 8.7 per cent. Inflation, another economic indicator, has also stayed at about 3 per cent since the mid-1970s (Gold, 1986). With this level of wealth and rate of growth, Taiwan will soon achieve the level of development at present enjoyed by the industrialized countries of Western Europe.

Taiwan has also done well compared to other countries. Although among Asian nations Japan, Singapore, and Hong Kong are wealthier, in terms of growth rate Taiwan shares top place with South Korea. Taiwan has also been remarkably successful in keeping down the rate of inflation. When

Table 3.1 Taiwan's Real GNP Growth Rate, Per Capita GNP, and Per Capita GNP Nominal Growth Rate, Selected Years

Year	Real GNP Growth Rate[a] (%)	Per Capita GNP[b] (US$)	Per Capita GNP Nominal Growth Rate[b] (%)
1952	8.3	196	35.2
1955	4.2	203	14.6
1960	3.1	154	17.1
1961	3.5	152	8.5
1962	4.7	162	6.8
1963	6.2	178	9.8
1964	9.1	203	13.7
1965	7.9	217	7.2
1966	6.1	237	9.0
1967	7.9	267	12.7
1968	6.6	304	13.7
1969	6.6	345	13.4
1970	9.0	389	12.8
1971	10.7	443	14.1
1972	11.3	522	17.8
1973	10.7	695	27.3
1974	−0.7	920	31.5
1975	2.5	964	4.8
1976	11.4	1,132	17.4
1977	8.1	1,301	14.9
1978	11.9	1,577	17.9
1979	6.4	1,920	18.6
1980	5.1	2,344	22.1
1981	3.8	2,699	16.3
1982	2.2	2,653	5.7
1983	6.9	2,823	9.0
1984	10.0	3,167	11.0
1985	4.1	3,297	4.7
1986	11.3	3,993	15.0
1987	10.7	5,275	11.2
1988	6.6	6,333	7.8
1989	6.2	7,512	9.5
1990	3.9	7,954	7.8
1991	6.2	8,815	10.4
1992	6.0	10,202	10.0
1993	5.9	10,566	9.6

Table 3.1 Cont.

Notes: [a] At 1986 constant prices before adjustment of terms of trade.
       [b] At current prices.
Sources: Council for Economic Planning and Development, *Taiwan Statistical Data Book 1992*, 30.
        Figures for 1992 and 1993 from Directorate-General of Budget, Accounting, and Statistics, *Quarterly National Economic Trends, Taiwan Area, Republic of China*, 64 (February 1994): 2–4.

compared over time, Taiwan also seems more consistent than the Western industrialized countries where such economic indicators as growth rate and inflation rate are concerned, even though Taiwan had its share of trouble during the oil crisis in the mid-1970s (Directorate-General of BAS, 1987b: 7). The only two industrialized countries that have been able to sustain this kind of economic performance are Japan and the former West Germany (Directorate-General of BAS, 1987b: 7).

Rapid economic growth has drastically changed Taiwan's social structure. In 1945, the majority of the people were peasants living in the countryside. By 1961, the agricultural population had fallen to 49 per cent of the total, and by 1986, it had dwindled to 22.1 per cent (Directorate-General of BAS, 1987b: 83). In terms of distribution of gross domestic product (GDP), in 1951, agriculture represented 32.47 per cent of GDP, industry 23.6 per cent, and services 43.67 per cent. By 1987, these percentages had changed to 5.27 per cent for agriculture, 51.99 per cent for industry, and 42.74 per cent for services (Directorate-General of BAS, 1988: 32). In 1993, the figures were 3.46 per cent for agriculture, 40.63 per cent for industry, and 55.91 per cent for services (Directorate-General of BAS, 1994: 9). In about thirty years, Taiwan had transformed itself from an agrarian to a modern industrial society.

As development theorists stress, 'no community can long expect to be self-governing unless the members of that community enjoy a minimum level of material well-being' (Cohen, 1971: 109). Economic prosperity opens the door to higher levels of education and greater exposure to the media, and enables members of a society to be informed about politics and political

Table 3.2 Per Capita GNP (1990), Average Annual Growth Rate of Per Capita GNP (1965–1990), Average Annual Rate of Inflation (1980–1990), GDP Growth Rate (1980–1990) in Selected Countries

Country	Per Capita GNP (US$)	Per Capita GNP Growth Rate (%)	Annual Rate of Inflation (%)	GDP Growth Rate (%)
India	350	1.9	7.9	5.3
China	370	5.8	5.8	9.5
Philippines	730	1.3	14.9	0.9
Thailand	1,420	4.4	3.4	7.6
Costa Rica	1,900	1.4	23.5	3.0
Malaysia	2,320	4.0	1.6	5.2
Mexico	2,490	2.8	70.3	1.0
Portugal	4,900	3.0	18.1	2.7
S. Korea	5,400	7.1	5.1	9.7
Saudi Arabia	7,050	2.6	-4.2	-1.8
Taiwan	7,954	7.1	3.4	7.7
Singapore	11,160	6.5	1.7	6.4
Hong Kong	11,490[a]	6.2	7.2	7.1
Belgium	15,540	2.6	4.4	2.0
United Kingdom	16,100	2.0	5.8	2.4
Australia	17,000	1.9	7.4	3.4
Netherlands	17,320	1.8	1.9	1.9

Table 3.2 Cont.

Country	Per Capita GNP (US$)	Per Capita GNP Growth Rate (%)	Annual Rate of Inflation (%)	GDP Growth Rate (%)
France	19,490	2.4	6.1	2.2
Canada	20,470	2.7	4.4	3.4
United States	21,790	1.7	3.7	3.4
Germany[b]	22,320	2.4	2.7	2.1
Norway	23,120	3.4	5.5	2.9
Sweden	23,660	1.9	7.4	2.2
Japan	25,430	4.1	1.5	4.1
Switzerland	32,680	1.4	3.7	2.2

Notes: [a] Refers to GDP data.
[b] Federal Republic of Germany before unification.

Sources: World Bank, *World Development Report 1992* (New York: Oxford University Press, 1992), 218–21.
Figures for Taiwan: Council for Economic Planning and Development, *Taiwan Statistical Data Book 1992*, 27, 30–1; Directorate-General of Budget, Accounting and Statistics, *Social Indicators in Taiwan Area 1991*, 368.

issues. The more developed a society is, the more pluralistic it becomes, and it also becomes more difficult for an authoritarian regime to govern effectively (Huntington, 1984: 199). The evidence for this argument is more than clear: most democratic countries are developed and industrialized, and democratization is taking place primarily in the newly industrialized countries of Southern Europe, Latin America, and East Asia. The stable liberal democracies tend to have much higher per capita GNPs than the less democratic countries.

It is, of course, difficult to ascertain at what level of economic development a country may be expected to see a movement towards democracy. But Taiwan, with a per capita GNP of US$10,566 in 1993, is certainly in the newly industrialized category. This level of prosperity is close to that of many West European countries in the 1970s or Japan and the United States in the 1960s, suggesting that with its current level of development, Taiwan already has a solid economic basis for democracy.

Income Distribution

The economic development policies of some Third World countries have been criticized by many as overemphasizing growth and underemphasizing distribution of wealth. Indeed, the excessive polarization of incomes has become one of the most persuasive excuses for some countries to stubbornly continue their centrally planned economies. Income polarization is also the most important source of social and political unrest, as the only possible way out for the very poor is to overthrow the political order established by the rich and powerful (Huntington, 1968: 56–9). But because of the seeming incompatibility of growth-oriented and wealth distribution-oriented policies, not many Third World countries have managed to achieve a high rate of growth while maintaining a good distribution of national wealth, or vice versa. This problem poses a serious dilemma for many countries in their attempts to modernize.

Taiwan, in contrast, has both a rapidly growing economy and a fairly even distribution of wealth (see Table 3.3). Moreover, as we saw above, inflation, which hurts lower-income families the most, is also kept low. Taiwan's income distribution

Table 3.3 Income Distribution in Selected Countries, Selected Years

Country	Year	Distribution[a]
1. Taiwan	1991	4.97
	1990	5.18
	1988	4.85
	1986	4.60
	1983	4.29
	1980	4.17
2. Japan	1991	4.60
3. Sweden	1981	4.61
4. India	1983	5.11
5. Netherlands	1983	5.55
6. S. Korea	1991	5.70
7. Israel	1979–80	5.79
8. Finland	1981	5.97
9. Italy	1986	6.03
10. Norway	1982	6.37
11. France	1979	6.48
12. United Kingdom	1979	6.81
13. Canada	1987	7.05
14. Switzerland	1982	8.58
15. Spain	1982–83	9.59
16. United States	1991	8.90
17. Australia	1985	9.59
18. Singapore	1991	9.60
19. Venezuela	1987	10.77
20. Malaysia	1987	11.13
21. Colombia	1988	13.25
22. Costa Rica	1986	16.51
23. Brazil	1983	26.08

Note: [a] Ratio between the income of the highest 20% of households and the lowest 20% (calculated by the author).

Sources: Figures for Taiwan, Japan, South Korea, the United States, and Singapore announced by the Directorate-General of Budget, Accounting, and Statistics on 18 March 1993, and reported in the *Independence Evening Post*, 18 March 1993, 3. Figures for other countries obtained from *World Bank, World Development Report 1992* (New York: Oxford University Press, 1992), 276–7.

compares well with that of the developed countries and even with that of welfare states such as Sweden and Norway, a very significant achievement indeed.

Some Taiwanese scholars see the chief characteristic of the recent changes in Taiwan's social structure as the rise of the middle class, which encompasses educators, managers, entrepreneurs, politicians, party cadres, civil servants, and middle- to high-ranking military personnel, who make up nearly 30 per cent of the population (Kao et al., 1988: 174–6). But according to statistics, the increase in income has been evenly distributed, and the whole of society has become affluent, not just the middle class.[10] One survey revealed that more than half of the population perceive themselves as belonging to the middle class (Kao et al., 1988: 176). This is a very good indication of the even increase of social wealth.

Observers of the Taiwan economy would point out that since the second half of the 1980s, the gap between rich and poor in Taiwan has widened. Because of the excessive amount of money available, many people invested heavily in the stock and real estate markets, forcing up stock prices and housing costs in the years from 1987 to 1989. These investors were able to reap huge short-term gains. However, those who could not afford this kind of investment became relatively worse off because of the escalating cost of housing in urban areas. It became almost impossible for middle- or lower-income urban wage earners to buy a condominium, the most common form of housing in Taiwan. The situation became so bad that a pressure group, 'Snails Without Shells' (*Wuke guaniu*), was formed to persuade the government to do something about the urban housing situation (*The Journalist*, 122).

Most democratic countries have a more egalitarian income distribution than those which fluctuate between authoritarian rule and periods of democracy. An egalitarian income distribution creates a society in which most people are likely to be satisfied with their economic status. It is also less likely to produce a polarized society or potentially violent class confrontation. In countries where national income is very unevenly distributed, as it is in many Latin American countries, there is great potential for violent confrontation since although the economy may be growing rapidly, the majority of the people have no share in it. In societies like these, only a

military dictatorship can contain such high levels of dissatisfaction among the masses. Taiwan's income distribution frees
it from a potentially deadly confrontation between rich and
poor or ruler and ruled, and provides a peaceful environment
for democratic transition to take place.

Social scientists often debate the virtues of economic indicators, such as per capita GNP and growth rate, because
they do not show the structure of national wealth. They would
prefer to examine other social indicators—for example, the
infant mortality rate, the number of physicians and hospital
beds per thousand population, the school enrollment rate, the
literacy rate, and the percentage of the population engaged in
agriculture and industry—to determine a country's level of
development. Where most of these indicators are concerned,
Taiwan is also ahead of the majority of its Asian neighbours
(see Table 3.4) (Directorate-General of BAS, 1987b: 426–9).

For example, the proportion of the population involved in
agriculture and industry indicates the degree of mobilization
and urbanization in a society. With less than 13 per cent of
its labour force working on the land, Taiwan has clearly
emerged from traditional society. Many people in Taiwan have
moved from isolated farming villages into large urban centers
where more employment opportunities can be found, leaving
the countryside to the older generation who farm the land
and often look after grandchildren whose parents work in the
cities.

Perhaps the most important aspect of Taiwan's social development is its high literacy rate and high education enrollment
rate. Being able to read and write means having access to
more abstract ideas and more information. A higher literacy
rate means that more people are capable of becoming politically informed through the print media. Educated people
are thus better able to make political decisions based on their
own understanding of politics. Majority rule is meaningful only
when the majority of the population are capable of making
their own choices and decisions on either candidates or issues.
In terms of the major social indicators, Taiwan is indeed
becoming a developed country, and it has a fairly advanced
social basis for democracy.

Another important indicator of socioeconomic development
is the availability of channels of communication. Government

Table 3.4 Major Social Indicators in Selected Countries, 1990

	A	B	C	D	E
Taiwan	85.4	34.1	92.4	12.8	40.9
Singapore	58	11.8	87.6	0.3	35.7
Hong Kong	66	13.1	88.1	0.9	36.0
Philippines	54	28.2	89.7	45.2	15.0
S. Korea	79	39.2	96.3	18.3	35.1
Japan	96	30.7	100.0	7.1	33.5
United Kingdom	79	23.5	100.0	2.1	28.6
W. Germany	85	33.7	100.0	3.8	40.1
Netherlands	82	32.4	100.0	4.5	25.9
United States	88	63.1	95.5	2.9	26.5

Notes: A: Secondary education enrollment as a percentage of 12–17-year-olds.
B: Higher education enrollment as a percentage of 20–24-year-olds.
C: Adult literacy as a percentage of population above age 15 years.
D: Agricultural labour force as a percentage of total labour force.
E: Industrial labour force as a percentage of total labour force.
Source: Directorate-General of Budget, Accounting, and Statistics, *Social Indicators in Taiwan Area 1991*, 366–9.

statistics in 1991 showed that 99 per cent of households in Taiwan owned a colour television set, compared to about 78 per cent in 1981. While the percentage of households subscribing to newspapers went up slightly from 59 per cent in 1981 to 62 per cent in 1992, the popularity of the telephone increased dramatically from 61 per cent to 94 per cent of households in the same period (Directorate-General of BAS, 1991: 22, 27, 394–5). All three indicators show that modern means of communication have been made available to almost every household in Taiwan, giving Taiwanese families access to up-to-date information.

Furthermore, in a survey conducted by National Chengchi University in 1989, more than 52 per cent of respondents said

Table 3.5 Media Exposure, Taiwan and the United States (percentages)

	Taiwan (N = 2647)	United States (N = 2257)
Read Papers	70.1	76.9
Read Papers 7days/week	52.1	39.1
Watch TV News	83.2	96.3
Watch TV News 7days/week	54.9	45.6

Sources: Taiwan: 1989 Taiwanese National Election Study, conducted by National Chengchi University for the Research, Development and Evaluation Commission, Executive Yuan, ROC. United States: 1984 American National Election Study, via Inter-University Consortium for Political & Social Research (ICPSR), University of Michigan.

they read a newspaper every day, about 70 per cent said they read newspapers regularly, and only about 26 per cent of respondents did not read newspapers. In the same survey, 54.9 per cent watched TV news every day, about 80 per cent watched regularly, and only about 11.5 per cent never watched TV news. When these results are compared with those of the 1984 American National Election Study survey, one can see that Taiwanese are exposed to approximately the same amount of news media as Americans (see Table 3.5). Although the quality of the media may be debated, there is no doubt that Taiwanese have good access to communications networks. This may have a profound effect on their political consciousness. Under such circumstances, government control through propaganda loses its appeal because people better understand their own interests. In other words, mass media exposure makes democracy more likely.

International exposure is also important to Taiwan's modernization. Foreign trade is one of the most important economic activities in Taiwan. Taiwanese businessmen are actively marketing their merchandise around the world, and in addition to financial rewards, they also bring back to the island an international perspective on society. Taiwanese students are seeking higher ·degrees abroad; US government statistics show that Taiwan was the primary source of foreign students in the United States in 1988 (*Central Daily News*, 29 December 1988: 1). The majority of Taiwanese students in the United States attend higher education institutions, and many of them return to Taiwan to teach at colleges or serve in the government. They bring back with them modern expertise, as well as new ideas and visions. The number of Taiwanese travelling abroad as tourists is also rising. According to government statistics, more than four million, or more than one in five, Taiwanese travelled abroad in 1992 (*Independence Evening Post*, 18 March 1993: 3). Contacts with the outside world have brought the isolated island of Taiwan into the modern world.

Sources of Economic Development

All the available social and economic indicators demonstrate that Taiwan is one of the most economically developed nations of the Third World. Economists would agree that aside from the legacy left by the Japanese colonial regime, which was discussed in Chapter 2, the most important source of Taiwan's prosperity is its hard working small- and medium-scale private firms. Out of 706,500 registered business enterprises in 1983 (and indications are that there have been no significant changes since that year), 98.6 per cent had an annual business revenue of less than NT$40 million, or the equivalent of US$1.1 million (*The Economist*, 5 March 1988: 11). These small private firms, usually family concerns, are highly productive and dynamic, and are easily reoriented to other types of production when the original products are no longer profitable. Such businesses are always on the lookout for new types of production. As a result, Taiwan probably has the highest number of business owners per head of population in the world.

The second factor affecting Taiwan's economic development is the Nationalist government's egalitarian economic policies in the 1950s. Taiwan's land reform of the early 1950s, for example, is recognized by many as one of the world's most successful such programs (Kuo, 1983; Gold, 1986; Huntington, 1968: 383). As a government transplanted from outside without extensive ties to the native population or any serious opposition to contend with, the KMT regime had more freedom than any other Third World government to carry out a large-scale land reform program (Gold, 1986: 65). Each stage of the program involved extensive redistribution of land, which is the most important source of social wealth in an agrarian society. The first step was to fix farm rents at 37.5 per cent of the crop yield (*sanqiwu jianzu*) to limit the landlords' share of the harvest. The second step was the release of public farmland (*gongdi fangling*) to increase the number of small-scale owner-cultivators. The third stage, land to the tiller (*gengzhe you qitian*), set limits on the area of land a landlord could own and redistributed the excess among the tenants (Gold, 1986: 65–6; Kuo, 1981: 48–55). As a consequence of land reform, the proportion of land tilled by owner-cultivators (*zigengnong*) increased from about 50 per cent to 75 per cent, the proportion cultivated by tenant farmers fell from 41.8 per cent to 16.3 per cent (Gold, 1986: 66), and the proportion of owner–cultivators in the agricultural population increased from 36 per cent to 60 per cent (Kuo, 1983: 27). The rural gentry was thus uprooted and a more egalitarian social structure was established in the countryside. Production patterns changed, productivity increased, and farmers were better off economically as a result (Kuo, 1983: 28–9).

Another of the KMT's egalitarian measures was its industrial nationalization policy. As soon as Taiwan was returned to the Nationalists, the government took over the wine, tobacco, and salt monopolies, coal mining, and all other Japanese-owned and operated industries. Electricity, petroleum, highways and railroads, and telecommunications were all made into government monopolies (Kerr, 1965). The government also owned major shares in most heavy industries, such as automobile manufacturing, transportation, glass, sugar, aluminium, cement, and so on, and regulations made it difficult for individual investors to compete with the government. Rice, the main staple and the most important crop in Taiwan, was managed by the government through the island-wide farmers' associations. Moreover, the KMT government also strictly controlled banking operations through its major shares in all banks. In other words, the very large public sector turned the government into the country's largest capitalist. In turn, the government reinvested the accumulated capital in public projects such as education, irrigation, highways, public transportation, and water supply, from which all citizens could benefit. In short, through these policies the government of Taiwan was able to keep the wealth of the country relatively well distributed and only a few businessmen such as Y. C. Wang (Formosa Plastics) and C. F. Koo (Taiwan Cement) were able to become wealthy capitalists (Gold, 1986: 71).

The third source of Taiwan's prosperity, and something which helped Taiwan to cross the threshold of economic development, was the policy shift from import-substitution to export-oriented industrialization in the 1960s. Lacking any ties with the local population, the Nationalist government needed an alternative source of regime legitimacy, and perhaps the only one available was economic performance. In the 1960s, in an effort to boost economic growth, the government eased controls on the import of raw materials and the export of finished products. Financial incentives, such as low-interest loans, were also introduced to encourage private investment and export activities (Kuo, 1981: 73–83). Taiwan's abundance of cheap skilled labour also attracted foreign capital which was invested in export-oriented industries. Taiwan's exports, especially to the United States, rose rapidly and allowed the government to amass large reserves of foreign exchange. In 1960, exports constituted only 11.5 per cent of expenditure on GNP (at current prices). By 1970, the proportion had risen to 30.4 per cent, by 1980 it was 52.6 per cent, it peaked in 1986 at 56.7 per cent, and by 1991 it was 47.1 per cent (Council for Economic Planning and Development, 1992: 44). This rate of increase indicates the rapid growth in Taiwan's export industries and the amount of wealth that was accumulated through export activities.

The Impact of Economic Development

Along with economic prosperity comes a change in people's political consciousness. As Huntington notes, the impact of social and economic modernization includes 'political mobilization [which] involves increased participation in politics by social groups throughout society' (Huntington, 1968: 34). In Taiwan, the most direct political impact of economic success was that the population began actively promoting its own political and economic interests in a variety of ways, including electing opposition politicians to the parliament (discussed

Table 3.6 Public Protest Activities in Taiwan, 1983–1987

	1983	1984	1985	1986	1987	Total	%
Political	5	4	20	35	106	170	11
Environmental	43	60	34	78	167	382	25
Livelihood	57	72	89	101	257	576	38
Labour Dispute	20	37	85	38	62	242	16
Other	18	9	14	19	83	143	9
Total	143	182	242	271	675	1513	[99]

Source: Hwang, 1988b: 6.

in detail in Chapter 4), and participating in protest activities and autonomous social groups and social movements.

Protest Activities

The most popular way in which people made their demands known to the government in the 1980s was through public protests, although not all protest activities were political in nature. Given the lack of formal and legal channels in Taiwan's institutional setup for making private interests heard (the government imposed harsh restrictions on freedom of speech and association), it was not unusual for private interests to be sacrificed in the name of the public good. But in the 1980s, more and more people were willing to challenge the establishment through collective protest activities such as rallies, marches, public hearings, demonstrations, blockades, petitions, and strikes.

Collective protest activities were on the rise throughout the 1980s (see Table 3.6), even before the lifting of martial law (Hwang, 1988a: 6). According to one government report, between 15 July 1987, the day martial law was lifted, and 31 March 1988, 1,408 collective protest activities took place in Taiwan, an average of five a day (*Central Daily News*, 20 April 1988: 2).

As mentioned above, the frequency of social protest activities began to rise in the early 1980s before the lifting of martial law, though the rise was more rapid after July 1987 when this strong psychological barrier was removed. But judging from the long-term trend, the number of social protests would have continued to rise even if martial law had remained in force, and this corresponds very well with modernization theory: the more a society is modernized, the more its people will make demands (Huntington, 1968: 32).

Hwang's study finds that there are four major categories of social protest in Taiwan: political, environmental, labour disputes, and protests affecting people's livelihood. Among the 1,334 protest activities listed in Hwang's study, 526 forced the targets to solve the problem immediately or promise to solve the problem later (Hwang, 1988b: 16). This shows that protests have become an effective means for promoting the interests of sections of the public. From this general trend, one can reasonably argue that the people of Taiwan are increasingly aware of their interests, and the demand for more political freedom and a greater degree of liberalization is part of this increased consciousness of their rights and welfare.

Even though the rise in protest activities is a national trend and there appear to be several social forces behind it, the diversity of the protests does not indicate a nationally co-ordinated or unified movement. Several voluntary social organizations were in operation during the 1980s, but most of the protests during that period were spontaneous in nature. The only protests that were apparently coordinated or directed were political ones, and those were usually directed by the Dangwai, the predecessor of the DPP. According to Hwang (1988b: 7), of the 170 political protests held between 1983 and 1987, the Dangwai was involved in 90, while the others were spontaneous. In other words, the Dangwai was the only nationally organized force consistently protesting against the government and demanding more democracy at that time.

To the government and the public alike, the most unexpected protest in the history of Taiwan was the one-day walk-out by railroad workers on Labour Day, 1 May 1988. The railroad is a government enterprise in Taiwan, and few expected that government workers would go on strike. But when the government failed to respond to the complaints of railroad workers concerning overwork, poor working conditions, and a lack of overtime pay, the workers simply took Labour Day off. The impact of the walk-out on commuters and other passengers forced the government to respond to the workers' complaints immediately, and working conditions and welfare were greatly improved. This event encouraged others who considered their legitimate complaints had been overlooked by their employers to take similar action.

Collective protest activities have an encouraging effect on the process of democratization in a transitional period because they reveal the deficiencies of laws and governmental institutions which no longer meet the needs of an increasingly complex society. The protest against the du Pont chemical plant in Lukang and other follow-up protests elsewhere indicated that the government had not institutionalized ways to predict public reaction to its policies. Under martial law such actions were thought to be out of the question. As a consequence, the government was completely unprepared to deal with problems of this kind and failed miserably. In addition, the Lukang protest exposed Taiwan's lack of environmental protection regulations. It was obvious that the government would have to review outdated or inadequate laws and regulations to keep up with the development of society. The protests of the mid-1980s apparently contributed to the termination of martial law. If the lifting of martial law is seen as a significant step towards democracy, these protests contributed to that end as well.

The protests forced the government to realize that it must make changes to accommodate the people's rising demands for their rights, as the cost of suppressing those demands had become too high. In other words, the government could no longer govern effectively by relying on traditional authoritarian means of control.

Political protests were by no means the most frequent form of protest at that time (Hwang, 1988b: 6), but they were certainly the most important to the political development of Taiwan. It is quite obvious that without an organized challenge to its authority, an authoritarian regime cannot be expected to relinquish power and undertake liberalization of its own accord. In the 1980s, political protests tended to become orchestrated events, rather than spontaneous riots such as occurred in Chungli in 1977 and Kaohsiung in 1979. They were used to expose the government's abuses of power and to provide the general public with a political perspective different from that of the government.

More significantly, in this period more people became willing to participate in rallies and demonstrations which, in essence, were demanding more freedom and democracy. The main participants in the demonstrations of the late 1970s and early 1980s were opposition politicians, their relatives, and hardcore supporters, all of whom had already put their careers and lives on the line. The demonstration that sparked the Kaohsiung incident, for example, attracted a crowd of only about one hundred (Peng, 1986: 74). After the mid-1980s, the magnitude of the demonstrations increased as the Dangwai acquired better organization and mobilization skills. For instance, the demonstration on Christmas Day 1987 attracted about ten thousand participants (*The Journalist*, 42: 13), and the one on 19 May 1989 gathered twenty-five thousand people (*China News*, 20 May 1989: 2). Participation obviously extended far beyond the close associates of opposition politicians.

The significance of the political protests is obvious: they challenged the very basis of the authoritarian regime and gradually eroded the legitimacy of authoritarianism in the eyes of the general public. The protests had a number of consistent aims: freedom of speech, the right to demonstrate, the right to form opposition parties, the lifting of martial law, and retirement of the pre-1949 parliamentarians. Faced with an onslaught of political protests and demonstrations, the government had no choice but to turn a blind eye to the formation of the DPP in September 1986 for fear of provoking a more violent reaction from the public. At this stage, martial law was in force in name alone; its termination only lifted the burden of public criticism off the government.

The protests also kept the issue of the pre-1949 parliamentarians in the public eye. And the fact that all three branches of the parliament were still packed with aging members did much to damage the government's legitimacy. The government was forced time and again to repeat the same old excuses for not retiring the senior members and holding a general election, but the longer the government held off, the stronger the pressure for change became. Public impatience on this issue was apparent in the results of the 1989 supplementary election for the Legislative Yuan. In addition to the twenty-one successful DPP candidates, many of the KMT's own candidates won public support for criticizing the government's foot-dragging on parliamentary reform.

Political protests did not usually achieve the same level of immediate success as that achieved by protest activities of other kinds (Hwang, 1988b: 16). But this is only to be expected since the issues involved were much more complex, usually being related to the structure of the political system, and change would have involved the decentralization of political power. Nevertheless, the political protests of the 1980s exposed the problems of the system to the public and made the KMT realize that changes were necessary if it was to maintain its legitimate claim to rule. In the long run, the political protests destroyed the basis of legitimacy of the authoritarian regime and forced it to implement political reforms that would satisfy the public. Ironically, the economic prosperity which the regime had worked so hard to achieve had created a population not content with the existing order. Socioeconomic change had gradually eroded the regime's basis of legitimacy, forcing it to give up single-party rule.

Juan Linz has argued that in a severely divided society, a politically influential military is almost inevitable (Linz, 1970: 267). In these circumstances, violent political protests might provoke a repressive reaction from the military and the conservative élite who are usually in alliance with each other. This argument seems plausible enough in view of the 1989 Tiananmen Square massacre in Beijing. However, political protests in Taiwan never reached the level of those in China in May 1989, nor did they threaten the KMT's dominant role in Taiwan's politics. Taiwan's political protests have never been as frequent or as violent as the student demonstrations in South Korea either. The conservative élite in Taiwan might have disliked the unceasing challenges to the KMT and they might have threatened to reimpose martial law (*The Journalist*, 64: 12–15), but the political protests did not constitute such a threat that a crackdown was inevitable. There is indeed a fine line political protests have to negotiate: they are necessary in order to push for reforms, but they must not be too strong or they will provoke a harsh conservative reaction.

Social Movements

Aside from the outbreaks of social protest discussed above, there also emerged in the 1980s a growing trend toward the formation of more general social movements, headed by a number of important voluntary social organizations. This development is a very important indicator of the pluralization of Taiwan society, in other words, the creation of a society that would not be so easily controlled and penetrated by the KMT. The most important of these movements involved students, environmental activists, labourers, consumers, and farmers.

Student Movements

University students in Taiwan have generally been more passive than their counterparts in South Korea, who form the backbone of political protests and are an important variable in their country's political development. Under martial law, student political activities were almost non-existent. But when Chiang promised to end martial law in 1986 this seemed to encourage students to seek more campus freedom and urge political reform.

The main reason for the lack of organized student political activities in Taiwan under martial law is usually viewed as threefold. First of all, the degree of ideological and behavioral control exercised by the school authorities was very high. Military officers were present on all university campuses to oversee student activities and enforce discipline, and students could be expelled for serious offences such as participating in anti-government activities. The KMT also had branches on every campus, and it was said that each class has its spy. Political courses such as the Three Principles of the People, Sun Yat-sen Thought, and mainland China studies were compulsory. A second factor is the tough competition students experience in order to get into college in the first place. Having gained a college place, most students preferred to indulge in more frivolous pursuits in their leisure time rather than get into serious debates on national issues or push for political reform. The third factor was the influence of parents who had gone through the terror of the 1940s and 1950s and who usually warned youngsters to stick to their books and stay out of politics. These three factors reinforced each other, and most students remained apolitical.

However, apparently encouraged by the formation of the DPP and the general relaxation of political control in the second half of the 1980s, students became more politically active. Student movements quickly spread from National Taiwan University in Taipei to National Chengkung University in Tainan and National Central University in Chungli. The activities ranged from sit-ins, demonstrations, hearings, and petitions, to the publication of underground journals. The students' demands were usually quite simple: the withdrawal of the military and the KMT from campus and the lifting of censorship on student publications (*The Journalist*, 21: 56–7). Students on some campuses also demanded a larger role in the school administration and decision-making. In a more extreme example, students at National Central University petitioned the Ministry of Education to have their president fired, as he had been appointed for his extensive ties with the ruling circle rather than for his academic reputation or administrative ability (*The Journalist*, 73: 78–9).

It was assumed by the public that the number of students involved in the movement was small, and that active participants were in a minority. This might have been true in most universities, but the lack of reliable data makes it difficult to support or refute these claims. Yet in some universities, challenges to the school authorities gathered quite phenomenal amounts of support. At the élite National Taiwan University, for example, a candidate frowned on by the school authorities was elected to head the student body (*The Journalist*, 67: 70), and at National Central University, only a very small minority of students supported the president (*The Journalist*, 73: 78–9).

The reasons for the rise of the student movement varied from campus to campus. At National Central University, it is quite easy to understand why so many students wanted to join the movement: the president had antagonized the students by being too authoritarian and refusing to listen to their demands (*The Journalist*, 73: 78–9). The situation at National Taiwan University was more complicated. Students on this campus frequently played a leading role in campus activities across the country. It was they who started the student movement in Taiwan, and they probably wanted to demonstrate that their movement was the most successful. Just like the students of Beijing University in China, National Taiwan University students wanted to become a model for the rest of the country to follow in the fight for campus freedom.

The student movement had only limited success. The Ministry of Education was forced to set limits on professors taking full-time government or party jobs in addition to their teaching responsibilities, the censorship powers of university authorities were curbed, and the role of the KMT and the military was reduced (*The Journalist*, 67: 70). Meanwhile, students seemed more willing to join off-campus protest activities. For example, on 20 May 1988, a group of students intervened in street fighting between police and demonstrating farmers in an effort to end the violence (*The Journalist*, 63: 52–3). Many students also volunteered to organize teams to help farmers whose livelihood was threatened by imports of American farm products. In sum, university students in Taiwan gradually became politically active members of society.

The Environmental Movement

Rapid economic development and industrialization without regard for the impact on the environment began to take its toll in the 1980s when pollution problems began threatening the health of tens of thousands of Taiwanese. Prevented by martial law from dealing with political issues, many writers in Taiwan turned their attention to environmental problems. But before the large-scale and prolonged protests against the du Pont chemical plant at Lukang in 1986, concern for the environment could hardly be termed a movement *per se*.

The year 1986 was a watershed for the environmental movement. In addition to the Lukang demonstrations, protests took place against the San Hwang chemical plant and the Lee Chang-yong plant elsewhere in Taiwan. However, the Lukang protests were the most important and had the most lasting effect on Taiwan's environmental movement. The local élite, KMT and opposition alike, jumped on the bandwagon and made the movement highly political. Not only did the protests result in the termination of the du Pont project, they also revealed the KMT's inability to deal with large social protests. It was the first time in Taiwan's history that an important economic project was disrupted through public pressure. Furthermore, the organizers of the Lukang protests turned themselves into professional protesters, helping to organize and train participants in other types of social movements (Chang, 1989: 55–7). Politicians interested in running for office vied for the support of the key organizers, and very few had the courage to speak out against the highly vocal environmentalists whatever their personal opinions on the subject. The movement continued to gain strength in the 1990s, and in one notable case, environmental activists were able to delay construction of Taiwan's fourth nuclear power plant at Kungliao, on the north coast, by lobbying politicians. Protests against the sixth naphtha cracker project at Litse in Ilan county forced Formosa Plastics to seek a site elsewhere. These successful efforts were modeled on the Lukang protests, and together they formed a nationwide movement for the protection of the environment.

The environmental movement continued to gather support throughout the island in the late-1980s and early-1990s partially because serious incidents of pollution continued to be reported. But there was also a financial motive. Some people were attracted to the idea of receiving compensation for the depredations of industry. Protests against the Lin Yuan Industrial Zone in September 1988 forced all the chemical plants in the zone to temporarily stop production, and the protests did not stop until the factory owners handed over NT$1.3 billion in compensation and promised to improve their pollution control. The large amount of compensation at Lin Yuan encouraged more and more protests of the same nature. In one example, in late April 1993, plants in the Ta She Petrochemical Zone were forced to stop operations and promise to compensate nearby villagers after repeated leakages of poisonous gas (*Liberty Times*, 3 May 1993: 4).

The environmental protests became a very important mechanism of public education after the mid-1980s, and debates between government planners and environmentalists received a lot of publicity. These incidents made people realize that something had to be done about the island's seriously damaged environment. As more and more people became aware of the issue, politicians at all levels were forced to pay more attention too.

The Labour Movement

Industrial workers as a social group are badly organized in Taiwan compared to those in other industrialized countries.

This situation was due to deliberate policies of the regime, which sought to maintain sociopolitical control and keep wages low so as to ensure the competitiveness of Taiwanese goods in the international market. According to Mau-kuei Michael Chang, a Taiwanese specialist on social movements, there was no such thing as a labour movement before the formation of the Legal Assistance Association for Labourers (*Laogong falu zhiyuanhui*) in May 1984 (Chang, 1989: 58). True, there were labour unions and associations in every fair-sized factory, but those unions were controlled, dominated, or closely monitored by cadres selected or appointed by the KMT (Chang, 1989: 61).

The result of the 1986 Legislative Yuan election, in which the KMT-dominated labour unions elected two relatively unknown DPP candidates and rejected the appointed head of the Chinese Federation of Labour, came as a shock to the government.[11] It was taken as a sign of workers' dissatisfaction with their working conditions and a weakening of the KMT's ability to monitor or control the attitude and behaviour of industrial workers. But apparently nothing was done, or could be done, by the KMT to reverse the situation. Labour unions began pressing for independence from government control, and politicians interested in winning the support of industrial workers also began to provide them with assistance and services. After martial law was abolished in July 1987, a DPP Legislator, Wang I-hsiung, and a group of union activists formed the Labour Party, anticipating that it would win the electoral support of Taiwan's five million industrial workers. Even though these hopes were not fulfilled, the party used its organizational strength to take an active part in various labour disputes, strikes, and protests.

The Consumer Movement

Consumer protection as a 'movement' is different from the other movements discussed above in the sense that it encompasses all members of society and its active participants did not use protests as a means to achieve their ends. Even though a national consumers association was formed as early as 1969 and a Taipei association in 1973, these organizations were unable to function effectively due to political control. It was not until the Consumers Foundation was founded in May 1980 that a consumer movement was formally launched. The foundation began by intervening in various food contamination cases. It carried out independent investigations, issued reports to the media, and reminded consumers to look out for defective and potentially dangerous products. Soon, the foundation began publishing a monthly, *Consumer Report*.

The Consumers Foundation was tolerated by the government under martial law mainly because it kept out of politics and distanced itself from the opposition. The foundation's success in maintaining its political neutrality also explains why it was so widely trusted and accepted by the general public.

The above discussion does not include all the social movements that emerged in Taiwan in the 1980s. Other examples include the women's movement and the movement to promote the rights of aborigines. However, those presented above certainly underline the growing involvement of Taiwanese in politics and social issues, and the rise of social movements corresponds very well with the generalizations of development theory. Taiwan's growing wealth, availability of education, and access to outside information are creating an ever more complex society; a society which demands that civil rights and individual freedom be respected and protected by a more democratic political system.

Summary Conclusion

Chiang Ching-kuo's decision to open up the political system in the mid-1980s is certainly the key to Taiwan's peaceful democratic transformation. The regime's willingness to loosen its control over society significantly eased tension between the government and the opposition and reduced the potential for violent confrontation. Chiang's decision was in line with the trend of Taiwan's social and economic development, which was making it more and more difficult for the KMT to rule in an authoritarian manner and to suppress the population's social and political demands.

As this brief survey of Taiwan's socioeconomic transformation has shown, Taiwan is an urbanized society with a mobilized and well-educated population that is both affluent and informed. International trade and cultural exchange with other developed and democratic countries has also introduced Taiwanese to societies in which citizens can make their voices heard through regular and fair elections for key decision-making posts. Under Taiwan's current socioeconomic conditions, the government is no longer immune from scrutiny by the news media, challenges from the opposition, and popular criticism. All these conditions form a strong basis for a successful democratic political system.

Chapter 4

■

Elections and Political Parties

POLITICAL parties are the single most important mechanism in liberal democracies for aggregating and representing different social and political interests in the political system. In democratic countries, people express their opinions and interests by voting for candidates representing different political parties in regularly held elections. The party which wins the support of the majority in a general election becomes the ruling party and acquires the power to make major policy decisions.

The subject of political parties and their relative strengths and weaknesses is crucial in assessing Taiwan's transition to democracy. The main concern of this chapter is to look at the major political parties in Taiwan and to analyse their electoral performance. As mentioned in Chapter 3, local elections have been held in Taiwan since 1948 and national elections since 1969, although the early national elections were on a very limited scale. Despite the fact that the authoritarian nature of the political system has meant that election results did not always determine national political outlook or identify the most important players in Taiwan's political arena, the elections that have been held so far provide some evidence for understanding the pattern of social support for different political forces.

In the early 1990s, only two of the sixty legally registered political parties in Taiwan carry any weight in politics (Copper, 1993). The only threat the KMT, Taiwan's largest party, has faced to its dominant position since 1949 was in the 1992 Legislative Yuan election; yet it still managed to win two-thirds of the total seats. In contrast, the DPP, formed out of a loose coalition of opposition politicians in September 1986, was able to send thirteen candidates to the legislature that year, twenty-one in 1989, and fifty-two in 1992.

KMT versus DPP in Elections

To understand the dynamics of elections in Taiwan one must first look at the unusual system which combines multi-member districts with the single non-transferable vote at both national and local levels. Western democracies generally employ either single member districts, as in Great Britain, France, and the United States, or a system of proportional representation, as found in some other countries on the European continent.

Though it differs dramatically from the systems found in Western democracies, Taiwan's electoral system is quite similar to that of Japan. It gives certain advantages to large, well-organized parties and there is often a great discrepancy in the outcome of an election between a party's share of the vote and its share of the seats. The KMT, like the Liberal Democratic Party in Japan, has been able to maintain majorities in the parliament and local assemblies chiefly because of the advantages offered by the electoral system.

The KMT has been overwhelmingly successful in elections at all levels (see Tables 4.1, 4.2, 4.3, and 4.4). Even after the 1992 election for the Legislative Yuan, which the DPP celebrated as a victory and the KMT recognized as a setback, the KMT still held a two-thirds majority in the chamber. However, the Dangwai and its successor, the DPP, have been able to win a sizeable share of the vote. By the early 1990s, the DPP had become the only political force with the potential to challenge the dominant position of the ruling party.

One reality one must bear in mind when studying electoral politics in Taiwan is that prior to the 1986 race for the Legislative Yuan, opposition politicians were forced to compete as individuals against the KMT party machine, being forbidden under martial law from forming their own party. Consequently, the results of pre-1986 elections are not an accurate measure of the relative strengths of the two parties. After 1986, the DPP, which inherited the traditional supporters of the Dangwai, became the only meaningful opposition party and the only party with the potential to replace the KMT in government.

Before 1969, the year of the first supplementary elections for the Legislative Yuan, the Taiwan Provincial Assembly was the highest representative body open to electoral competition. Provincial Assembly elections continued to be quite

Table 4.1 Comparison between the KMT and Dangwai/DPP in Provincial Assembly Elections, by Seats and Votes

Year	KMT			Dangwai (DPP in 1989)		
	Seats	Seats %	Vote %	Seats	Seats %	Vote %
1953	48	84.2	68.6	9	15.8	31.4
1957	53	80.3	67.8	13	19.7	32.2
1960	58	79.5	65.4	15	20.5	34.6
1964	61	82.4	68.0	13	17.6	32.0
1968	60	84.5	75.5	11	15.5	24.5
1972	58	79.5	68.9	15	20.5	31.1
1977	56	72.7	64.1	21	27.3	35.9
1981	59	76.6	70.3	18	23.4	29.7
1985	59	76.6	n.a.	17	22.1	n.a.
Average 1953–85[a]:		79.6	68.6		20.3	31.4
1989	56	72.7	60.15	16	20.8	22.5

Note: [a] Calculated by the author.
Sources: Seng, 1986: 12. Data for 1985: CEC, 1986: 10–11. Data for 1989: *China News*, 4 December 1989.

important after this date, however, as the provincial electorate is practically the same as that for national-level elections.

In Provincial Assembly elections prior to 1989, Dangwai candidates gained on average 31.4 per cent of the popular vote, but their share of seats was far lower than this. The opposite held true for the KMT. In the 1989 election, however, the first in which the DPP officially took part, the opposition's share of the seats was a much more accurate reflection of its share of the vote (see Table 4.1).

The same pattern of results can be found in the elections for county magistrates and city mayors, which are the only executive positions open to election. The KMT, again, has won most of the city and county contests, though opposition candidates have consistently been able to win a few seats. However, in terms of the share of the vote, opposition candidates have done much better (see Table 4.2). Again, the

Table 4.2 Comparison between the KMT and Dangwai/DPP
in County and City Mayor Elections, by Seats and Votes

Year	KMT			Dangwai (DPP since 1989)		
	Seats	Seats %	Vote %	Seats	Seats %	Vote %
1953	19	90.5	71.6	2	9.5	28.2
1957	20	95.2	65.0	1	4.8	35.0
1960	19	90.5	72.0	2	9.5	28.0
1964	17	81.0	73.1	4	19.0	26.9
1968	17	85.0	72.4	3	15.0	27.6
1972	20	100.0	78.6	0	0.0	21.4
1977	16	80.0	70.4	4	20.0	29.6
1981	15	78.9	59.4	4	21.1	40.6
1985	17	81.0	n.a.	4	19.0	n.a.
Average 1953–85[a]:		86.9	70.3		13.1	29.7
1989	14	66.7	53.5	6	28.6	37.59
1993	13	61.9	47.5	6	28.6	41.0

Note: [a] Calculated by the author.
Sources: Seng, 1986: 12. Data for 1985: CEC, 1986: 26–7. Data for
1989: *China News*, 4 December 1989. Data for 1993: *China
News*, 28 November 1993.

1989 election appears to have been a turning point for the
opposition, as together with independents they received about
45 per cent of the popular vote. In the 1993 election, the DPP's
share of the vote went up to 41 per cent (from about 38 per
cent in the 1989 contests), and popular support for the KMT
declined to less than half the vote.

The results of elections for the Legislative Yuan and the
National Assembly, the two chambers of the parliament that
are open to direct election, are found to exhibit a pattern
similar to those for the Provincial Assembly and county magistrates. The vote is distributed between the KMT and the
opposition on a ratio of about seven to three, but the ratio of
seats is about eight to two. Nevertheless, the opposition's percentage of the vote rose slowly in the 1980s and early 1990s
(see Tables 4.3 and 4.4), though its average share of seats did

Table 4.3 Comparison between the KMT and Dangwai/DPP
in Legislative Yuan Elections, by Seats and Votes
(Popularly Contested Seats Only)

Year	KMT			Dangwai (DPP since 1989)		
	Seats	Seats %	Vote %	Seats	Seats %	Vote %
1969	8	72.7	76.0	3	27.3	24.0
1972	22	78.6	70.2	6	21.4	29.8
1975	23	79.3	78.8	6	20.7	21.3
1980	41	78.8	72.1	11	21.2	27.9
1983	44	83.0	70.7	9	17.0	29.3
1986[a]	44	77.2	66.9	13	22.8	33.1
Average 1969–86[b]:	78.3	72.5			21.7	27.5
1989	72	71.3	58.5	21	20.8	27.2
1992	73	58.4	53.0	38	30.4	31.0

Notes: [a] DPP candidates were treated as non-partisans (*Dangwai*) in
official statistics for the 1986 election as the DPP was not
legalized until January 1989.
[b] Calculated by the author.
Sources: Seng, 1986: 12. Data for 1986: Compiled from CEC, 1987:
335–52. Data for 1989: *China News*, 4 December 1989. Data
for 1992: *China News*, 20 December 1992.

not rise in proportion. The results of the National Assembly
election of 1991 and Legislative Yuan election of 1992 show
the degree of support for the two main parties can vary quite
dramatically. While the DPP suffered a serious setback in 1991,
when it received less than a quarter of the popular vote and
won less than 20 per cent of the contested seats, it made a
comeback in 1992 to win more than 30 per cent of both contested seats and the vote. This enabled the party to form a
formidable bloc in the legislature.

Taiwan's election results reveal a slow but distinct rise in
support for the opposition which is becoming increasingly
able to reduce the discrepancy between its shares of votes and
seats. However, the KMT has still come nowhere near electoral defeat.

Table 4.4 Comparison between the KMT and Dangwai/DPP in National Assembly Elections, by Seats and Votes (Popularly Contested Seats Only)

Year	KMT			Dangwai (DPP since 1989)		
	Seats	Seats %	Vote %	Seats	Seats %	Vote %
1969	15	100.0	79.7	0	0.0	20.3
1972	27	75.0	72.0	9	25.0	28.0
1980	40	78.4	66.4	11	21.6	33.6
1986	46	78.0	64.8	13	22.0	35.2
Average 1969–86[a]:		82.9	70.7		17.2	29.3
1991	179	79.6	71.2	41	18.2	23.9

Note: See Table 4.3.
Sources: Seng, 1986: 12. Data for 1986: Compiled from CEC, 1987: 174–88. Data for 1991: *United Daily News*, 22 December 1991.

Strengths and Weaknesses of the Two Parties

The most important reason for the KMT's overwhelming electoral success is that it has been able to use its position as ruling party in an authoritarian system to discriminate against the opposition. There are other factors, of course, including the tactics of opposition politicians, the organizational strength of the KMT and its ability to manipulate local factions, and factional rivalries within the opposition camp. The importance of factional conflicts within the KMT cannot be overlooked either as they played a crucial role in the party's electoral setback in 1992 and are likely to remain an important factor affecting the outcome of elections in the foreseeable future.

The Nature of the Political System

Up until July 1987, the opposition's electoral performance was seriously hampered by the fact that under martial law, opposition politicians were prohibited from forming an organized force to compete with the ruling KMT. Oppositionists who criticized the government a little too fiercely or who attempted to form a political party often ran the risk of political persecution, while those who performed very well in elections or in office were liable to be co-opted into the system and given government posts, as exemplified by the cases of Henry Kao, a former mayor of Taipei, Su Nan-cheng, the former mayor of Tainan, Huang Shih-cheng, former Changhua county magistrate, and Chang Po-ya, former Chiayi mayor.

Caught between the threat of persecution and the temptations of co-option, the opposition also had to cope with a powerful mass media controlled or dominated by the government as part of the party-state structure. The three television channels were controlled by the provincial government, the military, and the KMT, and the radio stations were not able to operate without government interference or censorship. In the area of the printed media, government censorship was tight until martial law was lifted. Even then, the owners of the two most popular dailies, the *United Daily News* (*Lianhe bao*) and the *China Times* (*Zhongguo shibao*), were members of the KMT's Central Committee or Central Standing Committee. Opposition magazines were often banned for publishing anti-government views. It was not until Chiang Ching-kuo announced his decision to liberalize the polity that some newspapers began to carry reports critical of the government and sympathetic to the opposition. *The Journalist* (*Xin xinwen*), a weekly news magazine very popular among intellectuals that adopts a fairly neutral stand, began publication at this time. Throughout most of the martial law period, however, the opposition was almost totally silenced. The only opportunity its leaders could get to put their views across was during campaign meetings; thus their ability to gather substantial support was impaired.

Furthermore, the KMT's domination over the media enabled it to manufacture a public image as the only party that could ensure a stable future for Taiwan. Opposition politicians were frequently portrayed as a radical, violent, and irresponsible gang of conspirators. One possible reason why people voted for the KMT in such numbers was their fear of chaos should the opposition come to power. In a public opinion survey conducted by *The Journalist* immediately after Chiang Ching-kuo's death in January 1988, 39.6 per cent of respondents said that the opposition was the most serious obstacle to political development in Taiwan (*The Journalist*, 45: 17). This percentage seems very high compared to the 7.4 per cent who said that the biggest obstacle was Communist China, the 2.4 per cent who opted for the military, and the 15.1 per cent who chose the conservative faction of the KMT. The opposition in Taiwan had a serious image problem, and this was directly or indirectly caused by the KMT's control of the mass media.

Manipulation of the news media was probably most apparent during election campaigns. According to one authority, television news coverage of KMT candidates in the 1989 election was nearly twice as long as that of DPP candidates. More than three quarters of the time spent on KMT coverage consisted of positive messages, and almost none were negative. The coverage of the DPP, in contrast, was 60 per cent negative, and almost nothing positive was said (Liu, 1990: 54). Television news coverage became even more biased during the 1993 campaign. The KMT and its candidates took more than 80 per cent of the air time, leaving the DPP only about 10 per cent (*Liberty Times*, 24 November 1993: 17; *Independence Evening Post*, 10 November 1993: 2). It would be very difficult for the opposition to attract more voters in these circumstances.

Election campaigns are very expensive and sufficient financial support is crucial to any aspiring politician. The opposition's ability to raise funds from its supporters was hampered by the activities of the government security apparatus. Police harassment of opposition supporters often intimidated them into silence, and sizeable donations from either individuals or firms were virtually impossible to obtain because of fear of government persecution. This was still the case in the summer of 1988 when the author visited DPP headquarters to conduct interviews and was told of the case of a medical doctor whose clinic was 'inspected' by police just one day after he had made a donation to the DPP. The owner of the well-known Hai Pa Wang chain of seafood restaurants, who is a long-term financial supporter of the opposition, has his tax record watched very closely. Compared to the KMT, which owns and operates a large and profitable business empire, the DPP is very poor, and its candidates rely on American-style fund-raising activities. The KMT's budget for 1994 exceeds NT$5 billion, fifty times larger than that of the DPP (*Liberty Times*, 15 May 1993: 2).

The beginning of the 1990s, however, saw a remarkable relaxation of the general political atmosphere. The printed media became more pluralized and open, and opposition views could even be heard on television. One particularly meaningful development is the party political broadcasts that first appeared on television screens during the 1991 National Assembly campaign. Moreover, fear of political persecution was lifted when most political prisoners were set free and public discussion of sensitive political issues was permitted. But the 1991 National Assembly election, which the DPP admitted was a serious defeat, and the 1992 Legislative Yuan election, which the KMT acknowledged as a setback, showed that it was still difficult for the DPP to receive more than half of the popular vote. In other words, the authoritarian nature of the political system is not the only, nor the most important, factor determining the outcome of elections. One must look beyond the inherent bias in the political system to the strengths and weaknesses of the two political parties.

The DPP: Uneven Distribution of Votes

As previously mentioned, the Dangwai was a loose coalition of opposition politicians before the formation of the DPP in 1986. There was no national organization to work out a common, co-ordinated program, or to nominate candidates and conduct campaigns. As a result, some nationally known opposition leaders were able to win very large shares of the vote at the expense of other lesser-known opposition candidates.

The strength of the opposition lies in the ability of certain well-known politicians to capture a large share of the vote in elections (see Tables 4.5 and 4.6). For example, Ju Gau-jeng received 120,338 votes in the fourth district, in southwest Taiwan, in the 1986 Legislative Yuan election. The KMT candidate who came closest to him got only 93,028 votes. In the same year Hsu Jung-shu polled more than 191,000 votes from the third district, in central Taiwan, in her bid for the Legislative Yuan, leading the top KMT candidate by about 30,000 votes.

In the 1986 election, Ju and Hsu were not the only DPP candidates who gained high voter support. Hong Chi-chang received 161,384 votes in Taipei County for a National Assembly seat whereas the leading KMT candidate obtained only 115,068

Table 4.5 Highest Vote for the Opposition and KMT,
Minimum Vote Needed to Win, 1986 Legislative Yuan
Election, by District

Legislative Yuan District	Highest KMT Vote	Highest Opposition Vote	Minimum Vote to be Elected
1st District	109,979	159,374	78,021
2nd District	107,997	141,885	88,675
3rd District	130,634	191,840	86,910
4th District	92,485	120,338	72,864
5th District	118,806	112,433	94,209
6th District	55,426	n.a.[a]	47,503
Taipei City	139,617	134,839	74,288
Kaohsiung Cit	86,903	91,984	62,159
Fukien Province	16,833	n.a.[a]	16,833

Note: [a] Opposition candidates did not run in these districts.
Source: Compiled from CEC, 1987: 344–52.

votes. In Taipei City in the same election, 125,283 voted for
Chou Ching-yu, while the minimum number of votes needed
to be elected was about 63,000. Kang Ning-hsiang, the primary opposition leader after the 1979 crackdown, picked up
134,839 votes in his contest for the Legislative Yuan in Taipei
City in which the minimum vote needed to be elected was only
74,288. You Ching, one of the most charismatic DPP leaders,
received 159,374 votes in his bid for the Legislative Yuan in
Taipei County and the closest KMT candidate was about 50,000
votes behind. Hsu Kuo-tai received 141,885 votes in the
Legislative Yuan election in the northwest district, and he
beat the leading KMT candidate by approximately 34,000 votes.
In the Legislative Yuan election of 1986, the highest vote in
five of the nine districts was won by an opposition candidate,
and four of these won by a wide margin (see Table 4.5). While
the 1989 election for the Legislative Yuan followed the same
pattern as 1986 and a few popular opposition candidates were
able to win a large number of votes, the 1992 election is more
illuminating. The 1992 Legislative Yuan results once again
follow the same pattern, even though the electoral districts

Table 4.6 Highest Vote for the KMT and DPP, Minimum
Vote Needed to Win, 1992 Legislative Yuan Election, by
District

Legislative Yuan District[a]	Highest KMT Vote[b]	Highest DPP Vote	Minimum Vote to be Elected
Taipei 1st	129,019	97,376	21,060
Taipei 2nd	56,278	73,726	28,048
Taipei County	235,887	119,661	36,845
Hsinchu City	29,810	44,318	29,810
Hsinchu County	57,231	70,731	57,231
Changhua	75,969	91,234	59,596
Ilan	56,005	74,036[c]	56,005
Hualien	46,527	26,605	26,605
Taitung	35,657	21,277	35,657
Nantou	53,801	91,526	48,253
Yunlin	55,101	66,017	49,147
Taichung City	77,915	58,805	57,206
Taichung County	84,335	78,148	58,501
Miaoli	67,623	45,976	56,355
Pingtung	71,559	78,181	48,339
Tainan City	56,128	75,468	50,721
Tainan County	73,616	105,073	49,235
Taoyuan	110,570	71,079	50,279
Chiayi City	57,934	58,145	57,934
Chiayi County	57,175	63,612	53,905
Kaohsiung 1st	51,231	94,323	33,027
Kaohsiung 2nd	40,809	58,094	32,456
Kaohsiung County	90,552	104,729	40,252
Keelung	48,583	37,418	37,418
Penghu	20,641	4,644	20,614
Kinmen	10,926	557	10,926
Lienchiang	1,648	n.a.	1,648

Notes: [a] Districts were revised in 1989, with one for every county
and city except Taipei and Kaohsiung with two each.
[b] Including KMT candidates running without party sanction.
[c] Received by Chen Ding-nan, a non-partisan candidate strongly allied with the DPP. After he was elected, Chen joined
the DPP Legislative Yuan Caucus and formally joined the
DPP on 31 July 1993. See *China Times*, 1 August 1993: 2.
Source: China News, 20 December 1992: 11.

were made much smaller. For instance, in Taipei's first district, Chen Shui-bian and Frank Hsieh, two prominent DPP legislators, received roughly 97,000 and 83,000 votes, respectively, which was 30 per cent of the total votes in the district. In Taipei's second district, newcomers Shen Fu-hsiung and Chang Chun-hong received about 73,000 and 67,000 votes each, far surpassing other candidates. In Changhua county, Weng Chin-chu and the DPP's former chairman, Yao Chia-wen, received about 91,000 and 85,000 votes, leaving the closest KMT candidates at least 10,000 votes behind. In Tainan County, Su Huan-tzu, relatively unknown prior to the election, received 105,073 votes, while Kao Yu-jen, a member of the KMT's Central Standing Committee who received the most votes among the KMT candidates, was 32,000 votes behind. In Kaohsiung County, the DPP's Yu Cheng-hsien and You Hung together received about 208,000 votes, while the total number of votes cast in the district was only about 534,000.[1] The high popular vote for individual opposition candidates was transformed into a political challenge to the government party in the national legislative body.[2]

The popularity of these opposition politicians is often transformed into political pressure against the government in the parliament. But ironically, the ability of some opposition leaders to capture very high popular support does not work in the opposition's favour under the current system for elections to the National Assembly, Legislative Yuan, and Provincial Assembly. Under this system, Taiwan is divided into electoral districts based generally on county and provincial-level city boundaries.[3] The number of seats in each district is determined by the total number of contested seats and the size of the district's electorate. The problem is that a high concentration of votes for one or two key candidates of the same party in a multi-member district damages the party's overall performance in the district, as other candidates lose out—always assuming each party can only draw on a limited number of voters.

Table 4.7 contains a set of hypothetical results in a five-seat district which illustrates the DPP's weakness and the KMT's strength (this will be discussed in the next section) where the distribution of votes is concerned. The opposition could have won two seats if they had coordinated their campaign and persuaded their supporters to vote more effectively. The

Table 4.7 Vote Distribution and Electoral Outcome in a Hypothetical District with Five Contested Seats

	KMT Candidates		Dangwai Candidates	
	Vote %	Elected	Vote %	Elected
	15	Yes	25	Yes
	15	Yes	5	No
	15	Yes	2	No
	15	Yes		
	8	No		
Total	68	4 (80%)	32	1 (20%)

concentration of votes on one candidate destroyed the chances of the other two.

Had opposition candidates been better able to coordinate their campaigns and distribute resources, more of them could have been elected. Taking the 1992 Legislative Yuan election in Taipei's first district as an example again, had the two DPP high-fliers been able to arrange for some of their support to be transferred to other DPP candidates, all four DPP people in the district would have been elected. In fact, the total vote received by the four DPP candidates in the district was sufficient to have sent five DPP politicians into the legislature, more than half of the contested seats. But the concentration of votes on two candidates led to the defeat of the fourth, and the DPP won only three seats. The situation was almost the same in Kaohsiung County. The total number of votes received by Yu Cheng-hsien and You Hung was enough to send three DPP legislators, more than half of the seats available in the district, into the parliament. However, the realization that they would never be able to arrange an even share of the vote led the DPP to nominate only two candidates in Kaohsiung. The same thing happened in Changhua and Tainan where more DPP people could have been elected.

The DPP is well-aware of its limitations in this area, but it is not easy for the young party to improve its campaign coordination and resource distribution capability. An effort was made to establish a coordination mechanism as early as

1986. This did produce some results, as can be seen from the Legislative Yuan election during that year in Kaohsiung. Two opposition candidates divided the city in two for campaign purposes and refrained from crossing into each other's areas. The result of this was that both were elected (personal interviews, 20 July 1988; 11 August 1988). In the 1992 Legislative Yuan election the opposition was able to reduce the difference between seats and votes gained to an unprecedented 0.6 per cent, thus demonstrating that it has the potential to achieve even better results if better candidate coordination can be obtained.

Nevertheless, some candidates in the 1992 election, such as Fan Hsun-li, the unsuccessful DPP candidate in Taipei's first district, complained that the front-runners cared very little about the party's overall performance and had engaged in a fierce campaign to boost their personal popularity. Fan's bitterest complaint against Chen and Hsieh was that they played on the tendency among Taiwanese to be sympathetic to victims of the regime. Both of them had experienced defeat in the past and they competed with each other for the highest number of votes as a matter of personal pride. During the campaign, they constantly reminded their supporters that they might be liable to KMT persecution if they were not returned to office. As a consequence, Chen and Hsieh took 75 per cent of the total DPP vote. This example alone illustrates that the opposition still has a long way to go in devising an effective mechanism for effectively apportioning popular support among its candidates.

The KMT: Strength in Organization

In contrast to the DPP, which depends heavily on the strong showing of individual politicians, the KMT's real strength, in addition to its ability to obtain very high vote counts, lies in the discipline it imposes on individual candidates. Responsibility for handling election campaigns lies with the party's Department of Organization Affairs, which has been quite successful in controlling the nomination process and dividing the KMT's resources among candidates (Lei, et al., 1986: 125–30; Peng, 1987: 57–8).

Not surprisingly for a quasi-Leninist party, the KMT has a firmly established organizational network throughout society.

Through its teams and cells and with reference to past voting records, the Department of Organization Affairs is able to determine the number of votes it is likely to receive in every locale. Confident in its predictions, the department calculates the maximum number of KMT candidates that could win in each electoral district and then nominates that number of candidates. Each KMT candidate is then asked to confine his or her campaigning to one particular area so that, as far as possible, no one takes votes away from anybody else (Lei et al., 1986: 125–30; Seng, 1986). This practice has ensured that most KMT nominees are elected in the contests for the National Assembly, the Legislative Yuan, and the Provincial Assembly.

In urban areas, the KMT's method of apportioning votes among candidates in a district is to assign 'responsibility zones' to individual candidates. Usually, a KMT candidate's responsibility zone consists of several *li* (sub-division) in one or more *qu*.[4] Because the *li* chiefs are almost always KMT members and the direct superiors of the *lin* (neighbourhood) chiefs, the KMT goes through its local chapters to mobilize party members in the *li* and *lin*. The *li* and *lin* chiefs then have to serve as the candidate's contact with local residents, distributing campaign fliers and possibly gifts to potential voters. Often, *li* and *lin* chiefs will visit residents in their districts personally, trying to persuade them to support the candidate. Sometimes they will be accompanied by the candidate himself.[5]

The KMT's zoning system for urban election campaigns has met with conspicuous success. The KMT's candidates have performed exceptionally well in their own zones, certainly much better than they have in other zones (Liu, 1990: 96–9). Calculations based on data from Legislative Yuan elections in the 1980s presented in Liu (1990: 103–6), show that the average differences between percentages of votes received in the candidate's own zone and outside were 19.7 per cent in 1980, 20.6 in 1983, 18.21 in 1986, and 12.2 in 1989.

The KMT's campaign activities and mechanism for vote-sharing among candidates are very different in rural areas, as revealed by KMT cadres in charge of factional affairs (the heads of the 'first section' in a county branch). Local factions are the key to the KMT's election success in country districts where the society is held together by kinship ties. In rural electoral districts the KMT is not actively involved in recruiting and training newcomers to politics. Instead, the factions calculate the approximate number of votes they can get and ask the party to nominate candidates they are confident will win (personal interviews, 19 May 1993). During the campaign, the nominees organize their relationship networks, mostly through local opinion leaders and *cun* (village) and *lin* (neighbourhood) chiefs, the so-called *zhuangjiao*, or *tiau-a-ka* in Taiwanese.[6] Several days prior to the polling, the local contacts come up with lists of voters that they are able to mobilize or influence, either through interpersonal relations or through bribery (*maipiao*, or vote-buying), which they present to the candidate. Then, just before polling day, the candidates distribute the money for the bribes to their local contacts. The turnout in rural areas where most of the residents are non-political is as high as in urban areas, and the main reason for this is the mobilization power of the factions. Most important of all, the KMT has been able to capitalize on the factions through its patron-client relationship with them: the KMT allows faction leaders to fill political positions and operate profitable businesses at the same time in exchange for their loyalty to the party and the government (personal interviews, 19 May 1993).

Local faction leaders and candidates' *tiau-a-ka* frankly admit that in rural areas, no KMT candidate would be able to win if he or she did not engage in vote-buying. The author was told that the return rate, that is, the ratio of those bribed to those who actually vote for the candidate, is roughly three to one. In an election where a minimum of 50,000 votes is needed to win, a candidate must bribe more than 150,000 voters to guarantee victory. Usually a candidate will 'buy' twice as many votes or more for fear of losing the election. The price of a vote in the early 1990s has varied from NT$300 to NT$1,000, depending on the type of election and the intensity of the campaign. The effectiveness of vote-buying is said to have declined to some extent because most voters 'sell' their votes to more than one candidate and because of intense campaigning by opposition candidates. So KMT candidates are forced to buy even more votes in order to offset the lower return on their 'investment' (personal interviews, 19 July 1991). But all in all, vote-buying is still the easiest and most effective way for a KMT candidate to get elected in a rural area.

The practice of vote-buying has been bitterly criticized by the opposition as it both violates the election law and distorts the results of elections. Indeed, an election is meaning-less when large numbers of rural voters do not know and may never even have heard of the candidate for whom they are voting—they vote for him only because they are paid to do so. The quality of the candidate, his or her ability, experi-ence, and stand on various issues, are rarely concerns for the majority of rural voters. If one considers the fact that about one-third of KMT voters would not have voted at all if they had not been paid, the degree of popular support for the KMT does not seem so overwhelming after all. Since vote-buying has worked so well for the KMT in rural areas and cracking down on it would seriously jeopardize the party's ability to dominate elections, the KMT has little incentive to enforce the election law, despite protests from Taiwan's academic community. Nevertheless, one can still reasonably argue that, even without the bribery, the KMT would have the edge over the opposition in the countryside because its candidates have more extensive contacts in society, compared to the DPP which depends heavily on personal charisma and uncertain popular support. Factions are still the key to the KMT's electoral suc-cess outside the larger cities.

In the 1994 election for township executives and county council members, and the subsequent election for the coun-ty council speakers, vote-buying was so rampant that President Lee Teng-hui was forced to speak out against it. Encouraged by President Lee, the Ministry of Justice, headed by Ma Yin-jeou, decided to launch a crackdown on the speakers elec-tion. More than three hundred members of county councils, or about one-third of the total, were indicted, the majority of them being KMT members (*United Daily News*, 18 May 1994: 4). It is not certain whether the crackdown will have a significant impact on vote-buying in future elections, but it demonstrates to the public that the government has at last begun to do something about the problem, even if its efforts force vote-buying underground.

In addition to the KMT's patron-client relationship with local factions, local governments, farmers' associations, irrigation associations, fishermen's associations, schools, and police departments—all KMT-dominated institutions—are used as channels for mobilizing support during elections. The heads of these institutions are usually also leaders of the local KMT, and they are instructed by their party chapter to support certain candidates. The DPP can hardly hope to compete with the KMT in this kind of campaigning, but as more DPP candidates are elected county magistrates and they work to depoliticize the local government administrations and other institutions, the KMT's overwhelming superiority will gradually be eroded.

The KMT is firmly established in Taiwan society. With about two million card-carrying members in Taiwan, or roughly one-tenth of the population, the KMT has been able to organize its party to its electoral advantage. The decision-making structure is highly centralized, from the chairman and the CSC, down to local branches and teams. The chairman is the most important decision-maker, other key figures being the secretary-general and deputy secretaries-general, members of the Central Standing Committee, and the directors of the various departments. This centralized command system has been challenged to some extent by popularly elected representatives, particularly since the beginning of the transition to democracy, but the general structure of the party has remained intact.

The KMT's structure allows it to mobilize its teams and party cadres to support party nominees during election campaigns (Seng, 1986). Its control over its two million members enables it to pinpoint the exact location of sources of support. In contrast, the DPP depends heavily on popular opposition figures to attract large but unorganized numbers of supporters and it has no means of determining how many voters are likely to vote for any particular DPP candidate. This is leading some full-time party workers to rethink the DPP's organization. They think that if they adopt some of the KMT's methods they will be better able to locate support and distribute votes among their candidates.

Factionalism in the DPP

Factional conflict has been a serious problem for the opposition ever since it began to form a united front in the late 1970s, and it has on occasion been the cause of election losses (Peng, 1986). Factional politics very often determines policy positions, ideologies, and attitudes towards the future of Taiwan within the DPP. Though conflict between the two main factions has eased since Hsu Hsin-liang was elected chairman in 1991, the potential for future eruptions cannot be ignored.

The most important source of factional division within the opposition in Taiwan has been ideological and philosophical differences concerning how democracy is to be achieved. A second source was the leadership style of individual opposition politicians. The 1977 elections for the Provincial Assembly and county magistrates were an important step forward for the opposition. A record twenty-one opposition candidates were elected to the Assembly and four were elected county magistrates (Hwang, 1985: 50). The election campaign also saw the first spontaneous outburst of public anger since 1947, the riot in Chungli over election fraud. The small victory of 1977 encouraged some opposition leaders to take a more radical anti-government line, though others maintained a cautious attitude. Opposition politicians gradually formed themselves into two ideological camps, one which sought to 'reform the system' (*gaige tizhi*), led by Huang Hsin-chieh, and one which was content with 'reform within the system' (*tizhinei gaige*), led by Kang Ning-hsiang.

In 1979 each of these two groups launched its own magazine. *Formosa*, published by Huang Hsin-chieh, gained the most widespread support and was indeed the most important political magazine ever published by the opposition (Peng, 1986: 75). The magazine was the focus of the opposition's first major effort to develop an island-wide united front against the KMT. Those who participated in the united front included politicians from many different backgrounds and regions with very different styles and ideologies. In fact, the *Formosa* group was more like a grand coalition of individual oppositionists than a faction.

But because of its radical stand and interest in mass campaigns, *Formosa* was closed by the Taiwan Garrison Command after the 1979 Kaohsiung incident, having only published four issues. Kang Ning-hsiang's journal, *The Eighties*, survived however. Compared to *Formosa*, *The Eighties*, edited by Antonio Chiang, was moderate, but it was also very critical of the government's policies. The main difference between the two journals was that *Formosa* emphasized the power of the masses in promoting democracy, while *The Eighties* looked deeper into the problems of society and politics (Peng, 1986: 74–6). Kang and Chiang believed in arguing their case with the

Figure 4.1 The Evolution of the DPP's Factions

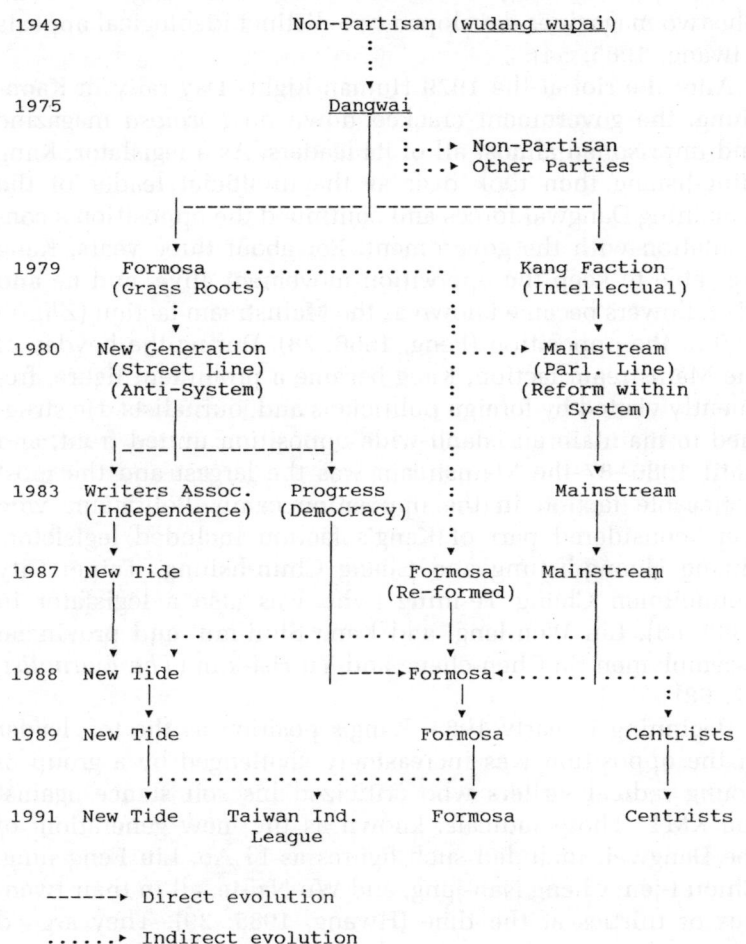

Sources: Peng, 1986; Hwang, 1985; personal interview, 12 August 1988; Hwang, 1992: 85.

government and achieving democracy through the ballot box— in all, a rather more intellectual approach to opposition than that adopted by the group around *Formosa* (Hwang, 1985: 52). The two magazines developed two distinct ideological appeals (Hwang, 1985: 54).

After the riot at the 1979 Human Rights Day rally in Kaoh- siung, the government cracked down on *Formosa* magazine and imprisoned almost all of its leaders. As a legislator, Kang Ning-hsiang then took over as the unofficial leader of the remaining Dangwai forces and continued the opposition's con- frontation with the government. For about three years, Kang was able to keep the opposition movement alive, and he and his followers became known as the Mainstream faction (*Zhuliu pai*) of the opposition (Peng, 1986: 78). During the heyday of the Mainstream faction, Kang became a prominent figure, fre- quently visited by foreign politicians and journalists. He strug- gled to maintain an island-wide opposition united front, and until 1986–87 the Mainstream was the largest and the most noticeable faction in the opposition camp. Politicians who were considered part of Kang's faction included legislators Huang Huang-hsiung and Chang Chun-hsiung, Taipei City Councilmen Chang Te-ming (who was also a legislator in 1980–83), Lin Wen-lang, and Kang Shui-mu, and provincial assemblymen Su Chen-chang and Yu Hsi-kun (*The Journalist*, 87: 68).

Beginning in early 1983, Kang's position as the top leader of the opposition was increasingly challenged by a group of young radical writers who criticized his soft stance against the KMT. These radicals, known as the 'new generation' of the Dangwai, included such figures as Li Ao, Liu Feng-sung, Chiou I-jen, Cheng Nan-jung, and Wu Nai-te, all in their twen- ties or thirties at the time (Hwang, 1985: 29). They argued that a mass campaign against the KMT government was essen- tial if Taiwan was ever to achieve democracy, and claimed in the new radical journal *Shen Geng* that Kang's call for peaceful competition with the KMT in the parliament was tantamount to caving in to the government. The opposition movement was now clearly divided into two ideological camps, one taking a 'parliament line' of peaceful electoral competi- tion and the other a 'street line' of mass confrontation. In 1983, the 'new generation' writers formed the Writers and Editors Association (Dangwai bianlianhui) as a united front against both the KMT and the moderate wing of the opposition. Later, the radicals launched their own journal, *New Tide* (*Xin Chaoliu*), from which their faction subsequently took its name. Most of these writers were not considered charismatic enough to run for election, but they provided a middle-level cadre for the opposition when it developed into an island-wide force and then a quasi-party. When the DPP was formed, members of the New Tide faction became its key cadres as they were the only professional organizers in the party.

As the government became more tolerant of the opposition in the mid-1980s, the New Tide faction turned towards the idea of a Taiwan independent from mainland China as the ultimate solution to the island's political problems, and they saw street demonstrations as the only way to overthrow the KMT and establish democracy. New Tide also stressed ideological purity—the measure of which was the degree of support for Taiwan independence. No compromise, cooperation, or negotiation with the government was to be sought, and there was to be no end to the struggle until the KMT was overthrown (personal interview, 13 July 1988). This attitude enabled the faction to develop into a tightly organized group, intolerant of other views within the opposition. The New Tide faction was criticized by some DPP leaders for limiting the appeal of the party, which looked likely to paint itself into a corner (personal interviews, 6 September 1988; 7 September 1988).

The Mainstream faction came under heavy attack from the radicals in 1983 as the next Legislative Yuan election approached, and the opposition camp soon realized that they would pay a heavy price at the ballot box for their factional wrangling. Indeed, Kang Ning-hsiang and two other Mainstream legislators lost their seats to the KMT (Hwang, 1985: 81). After his defeat, Kang seemed to modify his 'parliament line' somewhat, and in a personal interview, he stressed that the 'street line' was just as important (13 August 1988).

Meanwhile, the disbanded Formosa group reunited into a strong faction after Huang Hsin-chieh and Chang Chun-hong were released from prison on 30 May 1987. Compared to the New Tide, the Formosa faction was pragmatic in its outlook. Many of its members had held elected positions and were convinced that elections were just as important as street protests, if not more so. They also believed that democracy,

rather than independence, was what Taiwan needed at that
moment (personal interview, 6 September 1988). Each faction
rallied its supporters, and the power struggle between the two
threatened the unity of the newly established party (*The
Journalist*, 78: 19).

Thus the Formosa faction, the radical wing of the Dangwai
in the late 1970s, became a force for moderation and prag-
matism after 1987. The reason for this shift was firstly because
many members of Formosa were professional politicians who
saw elections as their only chance for achieving power. The
Kaohsiung incident had taught them that confrontations in
the streets would only provoke a KMT crackdown and ruin
their prospects for future election victories. Secondly, the New
Tide was chiefly made up of the most radical former members of Formosa, who formed the new faction after the Kaohsiung
incident. Formosa after 1987 was therefore much more moderate than it had been in 1979. Indeed, even in the 1970s,
Formosa only appeared radical in comparison to Kang's Mainstream faction. Now that its opponents within the opposition
were extremists rather than moderates, the Formosa faction
looked tame in comparison.

The power struggle within the DPP spilled over into the
Taiwanese community in the United States (*The Journalist*,
77: 39), and this was probably because factions in Taiwan
were searching for international allies to increase their power
and leverage within the party (*The Journalist*, 77: 26–9). The
most radical Taiwanese organization in the United States, one
which advocated terrorism and violent revolution against the
KMT, was the World United Formosans for Independence
(WUFI), and this organization was aligned with the New Tide
faction of the DPP. The Formosa faction, meanwhile, found a
natural ally in the Formosans' Association for Public Affairs
(FAPA), founded by the exiled dissident Peng Ming-min and
group of associates to lobby for the rights of Taiwanese
Americans. Unlike WUFI, FAPA advocated democracy through
electoral competition and self-determination.

The most noticeable of the smaller factions in the early
years of the DPP was that centered around Lin Cheng-chieh,
sometimes known as the Progress (*Qianjin*) faction. Lin had
been a member of the radical writers' group in the early 1980s,
but he broke with them in 1983 over the question of Taiwan
independence, and began to propagate his Green Party-style

opposition philosophy (personal interview, 13 August 1988) through his journal, *Progress Weekly* (*Qianjin*). Lin was known as an advocate of street confrontation with the authorities (his nickname was 'Little Master of the Streets'), but after Huang Hsin-chieh and Chang Chun-hung were released from jail he aligned himself with their reorganized Formosa faction and became one of the most vocal supporters of 'democracy now and independence later' (*Progress Weekly*, 13, 14, 15). However, Lin, a mainlander, could no longer stomach the DPP's increasingly pro-independence stance in the early 1990s and in June 1991 he withdrew from the party.

The election of a new DPP chairman in 1988 was the occasion of a showdown between the two largest factions, Formosa and New Tide, and between their two ideologies, democracy and independence. After months of bitter struggle and open conflict, Formosa won six of the eleven seats on the Central Standing Committee (CSC) and sixteen of the thirty-one seats on the Central Executive Committee (CEC). Huang Hsin-chieh, the leader of Formosa, was elected chairman of the DPP, with 123 votes against 97 gained by the incumbent Yao Chia-wen, a candidate favored by New Tide (*The Journalist*, 87: 57–9). The elections gave the more moderate Formosa a temporary edge over the more radical New Tide. One of the most important results of the leadership election was that Kang's Mainstream faction was not able to work out a coherent list of candidates for the CEC and was completely flushed out of the formal party establishment (*The Journalist*, 87: 68). Kang's followers were absorbed into the two main factions and the Mainstream faction vanished into history.

The two main factions clashed again in the next leadership election in 1991. This time, New Tide won seven CSC seats and seventeen seats on the CEC. But Formosa's candidate for chairman, Hsu Hsin-liang defeated Shih Ming-teh, the candidate favoured by New Tide, by a very small margin (*The Journalist*, 241: 57–8). This situation could have made it difficult for the party to make important decisions, but Hsu's pragmatic attitude and good relations with the leaders of New Tide prevented any serious stand-offs.

It was with factional conflict in mind that the DPP had specified in its constitution that the party should establish an independent and neutral arbitration committee to settle any serious disputes among its members. A few faction leaders even thought the party should learn something about the institutionalization of factional politics from the Liberal Democratic Party (LDP) of Japan (personal interviews, 6 September 1988; 11 August 1988). Their idea was that each faction should make public its program and ideology, and should name its followers, and that the composition of the CEC and CSC should depend on the number of delegates each faction sent to the party congress (personal interview, 13 July 1988). However, the idea of institutionalizing factions was not popular among the top DPP leadership, chiefly because it was viewed by many as a ploy by the better-organized New Tide to tighten its grip on the party.

Although some opposition leaders predicted that a final split in the party was inevitable (personal interview, 13 July 1988), the DPP was able to maintain basic unity, particularly in the run-up to major elections, because both sides realized that they must stay together if they were to have a chance of defeating the KMT (personal interviews, 13 July 1988; 6 September 1988; 7 September 1988). After Hsu Hsin-liang was elected chairman in October 1991, he tried to ease the conflict by appointing leaders from both factions to high positions in the party headquarters. In the most noticeable example, Chiang Peng-chien, the DPP's first chairman and an ally of New Tide, became secretary-general when Chang Chun-hong stepped down to run for the Legislative Yuan. During his election campaign, Hsu also encouraged Formosa to endorse the idea of Taiwan independence as the primary goal of the DPP (with the condition, suggested by Chen Shui-bian, that the goal should be reached through a plebiscite to avoid an all-out confrontation with the government) (*The Journalist*, 241: 41, 57–8).

It was also in 1991 that other centrist factions were formed, which effectively reduced tension between the two main factions. The Justice Alliance (*Zhengyi lianxian*) was formed by Chen Shui-bian, and his rival legislator, Frank Hsieh, formed the Taiwan Welfare State Alliance (*Taiwan fuliguo lianxian*). Each of these factions sought allies in the Legislative Yuan and the National Assembly. Though Chen was still considered close to Formosa and Hsieh was an ally of New Tide, the two alliances attracted a large number of DPP supporters who were dissatisfied with the way in which factional conflict had sapped the DPP's ability to defeat the KMT.

The return of radical members of WUFI, most notably Chang Tsan-hung, Li Ying-yuan, and Kuo Pei-hung, a new faction was formed that was represented in the legislature by Yen Chin-fu and Stella Chen. The WUFI faction was considered even more radical than New Tide, in the sense that the tight-knit clandestine group did not exclude the use of violent or revolutionary means in their efforts to establish a 'Republic of Taiwan'. Though the WUFI faction might be seen as a natural ally of New Tide, the home-grown radicals were actually hostile to the returned exiles who seemed to steal their independence thunder, not to mention poach their most radical supporters at the ballot box.

By 1992, factional politics in the DPP were looking more complicated but less confrontational than they had in the past. The factions were still united by the common goal of defeating the KMT and becoming the ruling party, and by the vague ideal of establishing an independent Taiwanese republic. At the same time, each faction had more foes to confront and more allies to align with, and no one group could dominates the party apparatus. In particular, the factions learned from the 1992 election that a united DPP would have a better chance of achieving electoral victory, especially at a time when the KMT was engulfed in an internal dispute.

The KMT's Internal Dispute

Compared to the DPP's very open and long-running factional conflict, the KMT's internal disputes have been more restrained—in the sense that the leaders of the different factions have usually not criticized each other openly—and relatively recent. The DPP's performance in the 1992 Legislative Yuan election, when it won nearly one-third of the seats, came as a shock to the KMT which had maintained an overwhelming majority in the national legislature. This setback should have come as no surprise, however, in view of the damage done to the party by the intense internal dispute of the early 1990s.

The KMT began to show signs of a split within the top leadership circle in December 1989, less than two years after Chiang Ching-kuo's death. President Lee Teng-hui's failure to decide on a running mate for the presidential election in March 1990 created tensions with those leaders who hoped to be

nominated, including the then Premier Lee Huan, Judicial
Yuan President Lin Yang-kang, head of the National Security
Council Wego Chiang (Chiang Kai-shek's second son), and
Defence Minister Hau Pei-tsun. Open conflict finally erupted
in February, when Lee Teng-hui announced his choice of Li
Yuan-zu, the secretary-general of the presidential office. Lee's
four opponents then decided to come up with their own
ticket for approval by the Central Committee, consisting of Lin
Yang-kang as presidential candidate and Wego Chiang as his
running mate. The conflict among the KMT's top decision-makers
rocked the entire nation and sent the stock exchange index
tumbling more than 6,000 points between 10 February and
25 March to its lowest level in three years (*United Daily News*,
11 February 1990: 6; 26 May 1990: 6).

In the midst of this power struggle within the ruling party,
Taiwan experienced its largest-ever student protests. The
students were disgusted by the fact that mainland-elected
members of the National Assembly would be electing Taiwan's
new president while ordinary citizens were excluded from
the electoral process. The demonstrations in Taipei, a kind of
small-scale Tiananmen, lasted from 16 March to 22 March,
the day after student representatives met Lee Teng-hui to press
their demands, which were the dissolution of the National
Assembly, the convention of a national affairs conference, the
abolition of the emergency constitutional provisions, and the
scheduling of democratic reforms (*The Journalist*, 159: 20–35).

Though Lin Yang-kang's decision to back out of the presidential race and Lee Teng-hui's nomination of Hau Pei-tsun
as the new premier defused what was Taiwan's worst political crisis, Lee's opponents kept up the pressure on the president. The party's Mainstream faction (*Zhuliu pai*), represented
by Lee Teng-hui and his top aide James Sung, and the Non-
mainstream faction, headed by Lee's four foes, could not work
out their differences, and each sought support in government
institutions including the chambers of the parliament. In the
Legislative Yuan, the Wisdom Coalition (*Jisi hui*), a group of
twenty-eight legislators, became the strongest supporter of the
Mainstream, and the eleven members of the New KMT Alliance
were the most outspoken challengers of President Lee. In the
National Assembly, KMT deputies also split along factional
lines over the constitutional issue of how to elect the president in the future. The mainstreamers argued that the

president of the ROC should be elected directly by the people of Taiwan, while the non-mainstreamers supported the old constitutional design of indirect election by the National Assembly. The debate became so intense that the KMT was forced to postpone a decision on the issue in order to avoid a split in the party.

In the 1992 Legislative Yuan election there were several electoral districts in which the two KMT factions were not able to agree on a list of candidates. Three members of the New KMT Alliance, Lee Sen-fong, Chou Chuan, and Chen Kuei-miao, were scratched from the nomination list, and Chen Che-nan, a member of the Wisdom Coalition, was expelled from the party for criticizing several non-mainstream leaders during the campaign. Lin Yu-hsiang and Wu Tzu, top leaders of the Wisdom Coalition, failed to win nomination in the party primaries. Jaw Shao-kong, head of the New KMT Alliance, and Wang Chien-shien, the recently resigned finance minister and Alliance ally, decided to run without party approval at the last minute. With the exception of Lee Sen-fong, all of these rebels ran in the election, most of them as independents, and succeeded in taking away many votes from official KMT nominees. For instance, Jaw Shao-kong ran in Taipei County and received more than 235,000 votes, the highest number of votes gained by a single candidate in the whole of Taiwan, and Wang Chien-shien received more than 129,000 votes in Taipei's first district (*China Times*, 20 December 1992: 11). Members of the KMT who ran as independents received a total of 710,000 votes, which was about 7.48 per cent of the total votes cast in the election (*China Times*, 20 December 1992: 6).

The worst effect of this rebellion for the KMT was the way in which many of these candidates fiercely criticized their party as part of their campaign strategy. The KMT's image as the ruling party was seriously damaged by this criticism, and many traditional KMT supporters decided to vote against the party as a result. The DPP's slogan of 'a united DPP against a divided KMT' touched many voters who were tired of the ongoing KMT factional dispute and wanted to have a stronger opposition party to check the ruling party. Others, dissatisfied with the party's reputation for corruption, which many of the rebels were eager to exploit, also switched their support to the party's opponents.

When the result of the 1992 election was announced, the KMT had more to worry about than the gains made by the DPP. The New KMT Alliance achieved a remarkable victory at the expense of the party in general. Under the leadership of John Kuan, the Non-mainstream faction was able to call on the support of about thirty KMT legislators and independents such as the former DPP member Lin Cheng-chieh and Oung Da-ming, the boss of the powerful Hualon group. These legislators were even prepared to enter into temporary alliance with the DPP to defeat their own party in several issue areas. Their attempt to topple the Mainstream faction's candidate for Legislative Yuan speaker failed, but on two other issues the Mainstream suffered serious defeats: a NT$30 billion project to move the Legislative Yuan to a new site was thrown out and the Office-holders' Personal Assets Disclosure Law (the so-called 'sunshine law') was endorsed by the legislature in defiance of the executive.

The DPP did not always co-operate with the KMT's Non-mainstream faction, however. The opposition is considered to be much closer to the Mainstream on the basic issue that government policies should be directed towards the shorter-term endeavour of promoting the welfare of Taiwanese rather than the distant goal of the unification of China. The non-mainstreamers are strong advocates of unification, whereas the more Taiwanese-oriented Mainstream faction is certainly not interested in unification at the expense of Taiwan's prosperity. The Mainstream and the DPP also agree that Taiwan should campaign to join the United Nations, not necessarily under the title 'Republic of China'. Consequently, if the Non-mainstream leaders were to question President Lee's adherence to the national goal of unification, the DPP would almost certainly come to his rescue.

This three-way coalition game between the DPP and the two KMT factions was a new departure in Taiwan politics. Though the KMT still had a two-thirds majority in the Legislative Yuan, many of its legislators could not be counted on to support the party leadership on all issues. This breakdown of party discipline meant that the KMT leadership often had to enter into delicate negotiations with both the opposition and its own factions in order to pass a bill.

Some KMT leaders, especially the younger legislators of the New KMT Alliance, frequently talked about splitting from the party. The alliance's strong showing in the 1992 election, following as it did on the failure of some of its incumbents to be nominated by the party, strengthened the faction's purpose. On 13 March 1993, members of the New KMT Alliance held a large rally in Taipei at which they announced their decision to form a 'political group,' something short of a new political party, and bitterly criticized the Mainstream faction, particularly President Lee Teng-hui (*China Times*, 14 March 1993: 2). On August 10, the alliance formally announced the formation of the New Party—its name no doubt inspired by Japan's New Party which had brought down the LDP—based on the seven core members of the New KMT Alliance, including Wang Chien-shien and Jaw Shao-kong. Although the New Party's chances of winning any more seats seem slim as things stand at the moment, their candidates may draw support away from the KMT in some districts and enable the DPP to make some gains.

The New KMT Alliance's slogan 'destroy the party, rebuild the party' (*huidang zaodang*) seems to have been put into practice, but it is difficult to estimate how much impact the split in the KMT will have on the party itself or the political system as a whole. Many leading members of the Non-mainstream faction hesitated to leave the KMT to which they had made a life-long commitment, and join the rebels. But some Mainstream leaders were quick to draw a comparison with Japan's Liberal Democratic Party which lost power in 1993 precisely because of such a split. Taiwan's political system is very different from that of Japan, where the party or coalition that wins a majority in the parliament automatically forms the government. Breaking the entrenched power of the KMT through elections is not an easy task for any alternative political force. It is especially difficult for a new organization which draws its support from the mainlander minority and young urban dwellers. Nevertheless, it is certain that the split will cost the KMT seats at national and local level and make the system more competitive. In the 1992 Legislative Yuan election, New KMT Alliance members won about 5.6 per cent of the total vote. Unless more KMT members defect to the New Party, its popular support is hardly likely to exceed that level, for the KMT will no longer 'distribute' votes to the rebels and it is likely to try to block their sources of support by nominating candidates from similar backgrounds to run against them. It will be a few years before the New Party can achieve the status of a recognized opposition party.

The Parties and Their Bases of Support

Having examined the electoral performance and internal problems of the KMT and the DPP, two important questions immediately arise. The first is, whom do the two parties represent? To answer this, we need to look at the background of the élites on both sides to see whether they consciously set out to represent any particular social group. The second question is, who supports the élites on the two sides? This is to ask whether they receive support from any particular social group, or whether any particular social groups seek to be represented by a particular party.

Élite Composition

KMT Élite

The KMT élite has undergone significant changes since Chiang Kai-shek died in 1975. Prior to Chiang's death, the party was, in both appearance and essence, a Soviet-style 'revolutionary democratic party'. In this austere atmosphere, the entire country, particularly the ruling party, was required to submit to sacrifice in order to fulfill the national goal of retaking the Chinese mainland and liberating it from communist control. A combination of military strength and ideological purity were essential for this purpose. The KMT's Central Standing Committee not surprisingly reflected this goal, being filled with military officers, party ideologues, propaganda specialists, and the heads of the intelligence services. Only a select few members of the Taiwanese élite, such as Huang Chao-chin and Shieh Tung-min, were admitted to the CSC.

However, changes were slowly but surely taking place in the 1970s, particularly after 1975, when Chiang Ching-kuo took over as the party chairman. The new chairman incorporated a number of reform-minded technocrats into the CSC and gradually transformed the KMT from a 'revolutionary' to a 'reformist' party (Chang, 1988: 3). It was also during this period that Chiang began to recruit more Taiwanese into the ruling circle. Tsai Hung-wen became a member of the CSC in 1976, and Lin Yang-kang and Lee Teng-hui in 1979. With new faces in the CSC, the party gradually became more concerned with economic performance and political development in the latter part of the Chiang Ching-kuo era, a marked shift from its previous obsession with ideology, national security, social stability, and political control (Cheng, 1989: 5). However, it was still quite obvious that all important policy decisions had to be cleared by Chiang Ching-kuo. He might consult with other members of the ruling circle, such as members of the cabinet and the CSC or his secretaries, when a policy decision was about to be made, but there is no doubt that his decision was final (*Central Daily News*, 12 January 1989: 4).

After Chiang's death, the KMT élite underwent another period of change. With Chiang in charge, the CSC had been more of a consultative body for the party chairman than the ruling party's highest decision-making body (*Independence Morning Post*, 15 July 1988: 2). But after the passing of the strongman, the CSC became the collective leadership of the party and the state. President Lee Teng-hui, who lacked seniority in the party establishment, was not able to lead in the Chiang style. Policy decisions, which are carried by majority vote in the CSC, could only be reached after intense debate and bargaining. Notable examples are the debate over the election of an acting party chairman and the issue of trade with the Soviet Union (*The Journalist*, 47: 9–13; 84: 12–15).

The composition of the current KMT ruling élite is revealing. After the party's Thirteenth Congress in 1988, a native Taiwanese (Lee Teng-hui) was formally elected chairman of the KMT, and a new CSC was appointed that was more than half Taiwanese in composition; no one could say that the KMT was a mainlanders' party anymore. The average age of the CSC in 1988 was 63.7, compared to 70.7 for the previous CSC, put together by Chiang in 1984. Of the thirty-one members, eight had earned Ph.D. degrees abroad (*Independence Morning Post*, 15 July 1988: 2). The Fourteenth Congress in 1993 went even further. Eighteen of the thirty-one members are Taiwanese and twelve have overseas doctorates. As younger and better educated recruits have replaced the old guard, the 'revolutionary' nature of the party has rapidly faded.

The backgrounds of CSC members since 1988 are evidence

of the KMT's new effort to link itself to Taiwan's changing society. In addition to the highest military, government, and party officials, the CSC now includes representatives from business circles, education, science and technology, veterans, overseas Chinese, local politicians, and labour (*Independence Morning Post*, 15 July 1988: 2). At the Fourteenth Congress, the party broke away from the tradition of an appointed CSC by reserving sixteen seats for election by the Central Committee. Of the fifteen members appointed by Lee Teng-hui, four were senior party leaders, ten were either cabinet ministers or heads of Yuan, and one was a top business leader. Among the elected members there were only two military officers, while the rest were legislators, chairmen of local assemblies, city mayors, and senior party officials (*United Daily News*, 22 August 1993: 2; 24 August 1993: 2). This arrangement is designed to demonstrate to the public that the KMT is willing to bring different voices from different sectors of society into its policy-making process.

There is a parallel trend towards change in the area of government administration. In particular, the cabinet reflects the KMT leadership's trust in younger, well-trained, and well-educated technocrats. The average age of the cabinet appointed after the KMT's Thirteenth Congress in July 1988 was fifty-eight, almost four years younger than its predecessor. Among the twenty-four cabinet members, eleven were Taiwanese and fifteen had overseas doctorates (*The Journalist*, 72: 23). Another cabinet was appointed in February 1993 after Hau Pei-tsun resigned under heavy pressure and Lien Chan, a Taiwanese himself, took over as the premier. The backgrounds of the cabinet members remained roughly the same, except that two university presidents were named to head the ministries of Defence and Communications (*China News*, 26 February 1993: 1). It is difficult to say whether these younger and highly educated members of the KMT élite are anxious for further democratization, but it is hardly likely that they will want to reverse the democratization process and return to the rigid polity of the past.

Changes in the élite's ethnic composition demonstrate the KMT's awareness of the need to bring the party into line with reality. The fact that the president/party chairman and over half of the CSC are Taiwanese has had a profound impact on the KMT's image as a mainlander party, and it is difficult for the opposition to argue any longer that Taiwan is ruled by a 'foreign' or 'colonial' regime (personal interview, 6 September 1988). As early as 1988, a leading opposition politician acknowledged that political conflict in Taiwan was no longer between Taiwanese and mainlanders; it was a competition between two Taiwanese parties (personal interview, 6 September 1988). A very significant proportion of Taiwanese dissidents in the United States, particularly those who belong to the FAPA, have also recognized that Taiwan's government and politics are no longer monopolized by mainlanders, and that overseas Taiwanese should abandon their revolutionary rhetoric against the Nationalist government and support peaceful reform instead (*The Journalist*, 84: 40–1).

Although the KMT has tried to balance the representation of mainlanders and Taiwanese in the party leadership and the government administration, the issue of 'provincial origin' still produces an undercurrent in society. What many Taiwanese complain about is that the proportion of government and party positions held by mainlanders—22.16 per cent of central and 25 per cent of local government civil servants, and 80.3 per cent of all government appointments—far exceeds the proportion of mainlanders in the population as a whole (*The Journalist*, 109: 41). However, some mainlanders, particularly those who have been part of the KMT hierarchy, fear that they are an 'endangered species' and that Taiwanese are swamping the KMT policy-making process. Indeed, the KMT power struggle of the late-1980s and early-1990s gave the public the impression that it was between Taiwanese (led by Lee Teng-hui) and mainlanders (represented by Hau Pei-tsun). Nevertheless, despite these factional differences, the KMT as a whole does not exclude either Taiwanese or mainlanders. It would hardly be possible for the KMT to maintain itself as the ruling party if it aimed to represent only one ethnic group, as it would soon earn the hostility of the rest of the population.

In the area of social class, the KMT also shows itself as a catch-all party and makes no attempt to orient itself towards any particular class or social stratum. Sun Yat-sen's Principle of People's Livelihood, had it ever been fully implemented, would have made the KMT a socialist party or a party representing workers and peasants, but the government's main economic strategy, outside the state monopoly of large industries, has been a neo-*laissez faire* market economy. The interests of workers and peasants have never been strongly articulated by the government, nor are they fairly represented in the KMT élite. Only one member of the KMT élite (Hsieh Shen-san, a legislator, CSC member, and since 1993 deputy secretary-general of the party) has a labour background. Not only does the KMT show no sign of representing a particular social stratum or group, it has actually tried hard to avoid the impression of exclusiveness, earning its reputation as a catch-all party.

The DPP Élite

In terms of background, the DPP élite is very different from that of the KMT. The DPP is a recent creation, unlike the KMT, which has been in power in Taiwan since the end of World War II. As a revolutionary party, the KMT has tried to demonstrate that it represents all sectors of society by including their representatives in the Central Standing Committee, the Central Committee, and the National Assembly. Many members of the KMT élite have achieved their current status through long years of service in the government, the military, or the party, rather than through elections. The DPP has never been in government and has no ties with the powerful military, so its power center does not contain any representatives of present or previous governments or the military apparatus. Most members of the DPP élite owe their positions to electoral success or to the contributions and sacrifices they made during the struggle for democracy.

The composition of the DPP's CSC and Central Executive Committee is also radically different from that of the ruling party's CSC in the sense that the majority of the DPP élite are current or former elected representatives, while most of the remainder are high level party cadres (Hwang, 1992: 80–1). Of the elected politicians, some have been lawyers, but most are professional political activists.

Another significant difference between the KMT and the DPP is that only a tiny minority of mainlanders have ever made it into the DPP's power center. After Fei Hsi-ping withdrew from the DPP in December 1988, the only notable mainlander among the DPP élite was Lin Cheng-chieh, the leader of the Progress faction and a member of the CEC. But since Lin left the party in June 1990, there has been no mainlander in either the CSC or the CEC. Even though the DPP has tried to avoid giving the impression that it excludes mainlanders, they can hardly be said to be well represented in the party center. However, the DPP's CEC and CSC are both elected by delegates to the party congress, so it is more difficult for the party's top leaders to balance their composition.

Despite this seeming lack of balance at the top, the DPP has always tried very hard not to alienate any social or ethnic group in its party platform. It may give the impression that it is a party of Taiwanese, but the DPP's platform has never implied that Taiwan should be ruled by Taiwanese only. The only article capable of offending any social group is the call to abolish the parliamentary seats reserved for overseas Chinese and the cabinet-level Mongolian and Tibetan Affairs Commission. And the only article which seems to be appealing to a specific social group is that advocating self-rule for aborigines (DPP Constitution and Platform: 17, 37). But overseas Chinese, Mongolians, and Tibetans are insignificant forces in Taiwan's electoral politics and aborigines represent only a very small minority (see Chapter 2).

Though the DPP as a whole tries to maintain an impartial stance, its more radical New Tide faction is determined to cultivate the support of farmers, workers, and college students, the so-called deprived social groups.[7] These groups are traditional supporters of the KMT, but their support is not always very consistent or solid. They are not mobilized to press for their own welfare neither are their interests articulated in public. In 1987, the average income of farmers was only about three-fourths the national average (Directorate-General of BAS, 1987a: 42), and they are always the first to suffer the effects of the trade wars with the United States (*The Journalist*, 64: 62). Working conditions for manual labourers in Taiwan are in many respects poor and their welfare is often ignored or sacrificed in the name of international competitiveness. College students, usually viewed as a privileged class in society, are often either co-opted by the KMT or coerced into staying out of politics by the presence of the military on campus. Without any political organization to articulate their interests and awaken their political consciousness, these groups are not likely to switch their support from the KMT to other parties.

New Tide's articulation of these groups' interests can be interpreted partly as an effort by the faction to portray itself as the champion of social justice for undermobilized and underprivileged social groups. A better explanation yet is that the faction is trying to broaden its basis of support in society. The more moderate Formosa faction, on the other hand, has tried to broaden its support by dissociating itself from any kind of socialism for fear of alienating the urban middle class and business interests. As a result of this conflict, it is ironical to see New Tide leaders involved in virtually every social movement and mass campaign on behalf of farmers, workers, and students, while the DPP as a whole maintains that the party has nothing to do with such agitation in order to clear itself of the charge of creating social unrest (*The Journalist*, 63: 54).

In summary, it is fair to say that the DPP has tried to be a catch-all party to compete against the catch-all KMT. Indeed, it would be difficult for the DPP to be anything else, since it is basically a coalition united by opposition to the ruling party rather than a centralized organization led by an ideologically unified élite. The attempt by the radical faction to mobilize workers, peasants, and students may or may not broaden the DPP's basis of support, but it is not likely to transform the DPP into a socialist party or one representing the underprivileged.

Mass Support

Support for the two main parties in Taiwan is, in fact, quite confused. No social group is overwhelmingly committed to supporting either of the parties, and neither party relies for its support on any specific group or class. In a sense, this is only to be expected when both parties are trying to attract all sectors of society, and this pattern of support also reinforces the catch-all nature of the parties. Neither party knows exactly where its main support is coming from, nor do they want to alienate any social group or class.

The KMT had its origins as a mass party mobilized to overthrow the Qing Dynasty in 1911, and its ability to mobilize the masses was greatly enhanced when Chiang Kai-shek reorganized it along Leninist lines in the 1920s. When the Nationalist government retreated to Taiwan in 1949, the entire party structure was kept largely intact, enabling the KMT to penetrate every corner of Taiwanese society through its well-established network. Despite the party's potential for communist-style mass mobilization, the KMT actually tried to demobilize the population and depoliticize political and social issues after the transfer to Taiwan, as there was no significant opposition capable of threatening its dominant position in politics and society.

In contrast to the KMT, the DPP has been very serious about mass campaigns and protest activities ever since its establishment in 1986. Often these activities take the form of rallies or speech forums conducted by prominent or charismatic opposition leaders, and also marches and demonstrations. As a result, the DPP has established itself as Taiwan's only genuine mass mobilization party, prompting the KMT to increase its mass activities in response. Since 1987, KMT legislators have sponsored their own outdoor gatherings and protest activities (*The Journalist*, 22: 51; *Central Daily News*, 6 March 1988: 2), though on a much smaller scale and much less frequently.

Public response to both parties' mobilization efforts has been unenthusiastic, as might be expected. A small fraction of the population is always more enthusiastic than the rest. But there is no sign that one specific social group or class is more likely to participate than others. For example, though Nationalist army veterans from mainland China are considered among the most loyal supporters of the government, a splinter group of veterans has taken a very visible role in the DPP's protest activities. In another example, farmers are also traditional KMT supporters, but the DPP's efforts to organize 'farmers' power associations' (*nongquanhui*) have gathered some strong support in central Taiwan. To the amazement of the KMT, a protest organized by these associations on 20 May 1988 turned into the worst riot in Taiwan since 1947 (*The Journalist*, 63), and 20 May has since become a traditional day of protest for farmers. In a final example, an increasing number of college students have become strong advocates of the DPP's political and social reform proposals, despite the fact that students' long-term lack of interest in politics had made them passive supporters of the regime. In short, there is no clear social or class divide between the two parties.

Data from opinion polls and the 1986 and 1989 Taiwan Election Surveys conducted by National Chengchi University for the National Science Council clearly indicate the precarious nature of the social support for the two parties. The following section contains an analysis of party support along socioeconomic lines, including income, social status, education, and provincial origin.

Income and Social Status

The middle class, no matter whether it is defined by income group or according to people's own perceptions, is seen by many authorities as the basis of Taiwan's political and social stability. This class is generally satisfied with a government which it sees as being responsible for its relative affluence and which it perceives as capable of preserving Taiwan's prosperity. However, one might also argue that the middle class would be enlightened enough to support the opposition in its struggle for more political freedom and a better political system. There is no indication which argument better reflects reality.

According to the 1986 election survey, the majority of those identifying with either the KMT or the opposition are from middle income families, but the Pearson's R is a negligible −.007 (see Table 4.8). The only difference is that those identifying with the opposition are drawn somewhat more heavily from the middle income population. However, reading the column percentages, the two parties' identifiers do not vary significantly enough to indicate that any income group gives either side more substantial support.

A cross-tabulation between party affiliation and subjective social class in the 1989 election produces similar results (see Table 4.9), this time with a negligible Pearson's R of .07.[8] These results basically confirm those of the 1986 survey that social status does not correlate well with party identification. Nevertheless, in both the 1986 and 1989 surveys, the lower income group and those who perceive themselves as having a lower social status are slightly more likely to identify with the opposition.

Education

Theoretically, one can argue both ways about the relationship between education and party identification in Taiwan. On the one hand, those who have benefited most from the Nationalist government's education policy are more likely to identify with

Table 4.8 Party Identification by Family Income, 1986 (percentages)

	Family Income[a]			
	Low	Middle	High	N
Dangwai	24.1 (9.3)	62.1 (14.9)	13.8 (16.0)	29
KMT	35.4 (90.7)	53.6 (85.1)	10.9 (84.0)	192
N	75	121	25	Total=221

Note: [a] Low: less than NT$20,000 per month; Middle: NT$20–50,000; High: more than NT$50,000. The US$ exchange rate in 1986 was NT$38=US$1.

Source: Based on 1986 National Election Survey conducted by National Chengchi University, courtesy of the Election Research Center, National Chengchi University, Taipei.

Table 4.9 Party Identification by Subjective Social Status, 1989 (percentages)

	Social Status			
	Low	Middle	High	N
DPP	23.9 (22.1)	68.1 (12.6)	8.0 (14.7)	138
KMT	14.0 (77.9)	78.3 (87.4)	7.7 (85.3)	830
N	149	744	75	Total=968

Source: Based on 1989 National Election Survey conducted by National Chengchi University, courtesy of the Election Research Center, National Chengchi University, Taipei.

the KMT, while on the other hand, the more educated a person is the more likely he or she is to have the necessary information to decide to oppose the government. The rather inconclusive data in Table 4.10 tend to reinforce this.

The cross-tabulation produces a Pearson's R of $-.008$ for the 1986 data and .09 for the 1989 data, not very different from that of the previous table. However, a slight relationship between education level and identification with the KMT can be detected. Comparing the row percentages, one finds that those who identify with the opposition are slightly more likely to have a lower level of education while those who identify with the KMT tend to have a higher level. Comparing the column percentages, one also finds that while less educated people are slightly more likely to side with the opposition, those with a higher education level are siding more with the KMT. But this relationship should not be stressed too much because of the very low correlation coefficients.

Provincial Origin

Provincial origin (Taiwanese or mainlander) could conceivably be a very significant variable that correlates with party identification, as no matter how much the KMT has tried to blend in with the local population and recruit the local élite, mainlanders remain very powerful in the government, the party, and the military. Moreover, from the mainlanders' point of view, the KMT may seem to be the only party that can protect their privileges and in which they can find powerful figures with whom to identify. The DPP has tried to shake off its image of an anti-mainlander party, even forming a Mainlander Association for Taiwan Independence (*Waishengren Taiwan Duli Cujinhui*), but the fact remains that one of the loudest arguments in the opposition camp is that Taiwan should be ruled by people born in Taiwan, not mainlanders (personal interviews, 13 July 1988; 8 August 1988), and support for the DPP among mainlanders is scarce and sporadic. The data in Table 4.11 essentially confirms this assumption.

The cross-tabulation produces Pearson's R correlation coefficients of .21 for 1986 and .18 for 1989, providing evidence that provincial origin does have a significant relationship with party identification. From the row percentages, one finds that those who identify with both the KMT and the opposition are

Table 4.10 Party Identification by Education, 1986 and 1989 (percentages)

		Education[a]			
		Low	Middle	High	N
1986	Dangwai	55.6 (13.0)	41.7 (11.8)	2.8 (2.8)	36
	KMT	47.7 (80.7)	39.9 (88.2)	12.5 (97.2)	281
	N	154	127	36	Total =317
1989	DPP	58.6 (17.2)	31.6 (9.4)	9.9 (11.1)	152
	KMT	42.2 (82.8)	45.9 (90.6)	11.9 (88.9)	1011
	N	516	512	135	Total =1163

Note: [a] Low: uneducated, primary, or junior high school; Middle: senior high or professional school; High: college and above.
Source: See Tables 4.8, 4.9.

drawn heavily from among Taiwanese, which is not surprising since about 85 per cent of Taiwan's population are Taiwanese. Nevertheless, the most important result is that those who identify with the opposition are almost exclusively Taiwanese. This trend is quite obvious from the column percentages as well. While a segment of Taiwanese support the opposition, almost all mainlanders identify with the KMT. Furthermore, the proportion of mainlanders identifying with the KMT exceeds that of mainlanders in the population as a whole. However, although one may conclude that the opposition is a Taiwanese party, there is no way one can call the KMT a mainlanders' party, because about two-thirds of KMT identifiers are also Taiwanese.

Provincial origin is indeed the only demographic variable

Table 4.11 Party Identification by Provincial Origin, 1986 and 1989 (percentages)

| | | Provincial Origin | | N |
		Taiwanese	Mainlanders	
1986	Dangwai	97.2 (15.6)	2.87 (1.08)	36
	KMT	66.5 (84.4)	33.5 (99.0)	284
N		224	96	Total = 320
1989	DPP	97.4 (16.2)	2.6 (1.6)	152
	KMT	75.2 (83.8)	24.8 (98.4)	1016
N		912	256	Total = 1168

Source: See Tables 4.8, 4.9.

that has a significant relationship with party identification. For example, the age variable could only achieve an R correlation of .09 for both the 1986 and 1989 surveys,[9] and, like income and education, it cannot be treated as an important factor in determining party identification.

Summary Conclusion

Political parties, including the party system in general and the individual parties' electoral strategies, electoral performance, and relations with the electorate, are of utmost importance in understanding how democracy is institutionalized in a given country. One important distinction between democratic and non-democratic countries is that in democracies, meaningful elections are held at a regular intervals to allow the general public to choose a ruling party. In democracies, the results of an election provide the legal basis for the winning party to form a government and thereafter implement policies that affect the course of national development.

The KMT, on account of its many years in power, is an overwhelming force in Taiwan politics. Since the establishment of the DPP in 1986, the opposition has become better organized and an admittedly somewhat skewed two-party system has evolved. Although the DPP has been able to offer a moderate challenge to the KMT, it has not been able to threaten the dominant position of the ruling party.

The KMT has always been decisively victorious in elections, and due to its superior organization and ability to manipulate the system has managed to win a greater share of seats than its share of the vote. In contrast, the opposition is less wealthy, less well-organized, and has, in the past at least, been plagued by factional conflict. The importance of party unity can be seen from the way that factional strife in the KMT took a toll on the party's performance in the 1992 election and is likely to provide a more united DPP with an advantage in future contests. Indeed, now that the KMT rebels have actually split from the party, there is a distinct possibility that the DPP will become the ruling party in the not too distant future.

A study of Taiwan election surveys reveals that neither the KMT nor the DPP relies entirely on the support of any particular social group or class. Nor do the parties target any specific ideological orientation—in the Western sense of left/right or liberal/conservative—in their campaigns. In fact, both can be described as catch-all parties. Analysis of survey data also indicates that party identification does not correlate well with income, education, or age, the most commonly utilized demographic variables. However, statistics do show that the 'provincial origin' variable, an ethnic factor present in few other countries, produces a rather significant correlation with party identification. Whereas both the KMT and the opposition depend heavily on the support of Taiwanese—the majority of the population—mainlanders almost without exception give their support to the KMT. This does not make the KMT a mainlander party, since two-thirds of those who identify with the KMT are Taiwanese, but there is no doubt that the opposition is in essence a Taiwanese party.

Chapter 5

A Clash of Views: The Political Élite on Key Issues

ONE of the most important factors affecting Taiwan's democratization, and one which has never before been systematically studied, is the role played by the élite in the transition process. Even though the concept of an élite is itself still a subject of academic debate, the national political élite, by definition, consists of individuals who are in strategic national policy-making positions or who have significant influence in organizations large or powerful enough to regularly or substantially affect national political outcomes (Higley et al., 1992: 8). To a large degree, it is members of the élite who decide when, at what pace, and to what degree an authoritarian political structure is to be transformed; what kind of political system will replace it; and how much power they are willing to give up in order to reach a binding consensus on a democratic constitutional framework. Needless to say, it is essential to examine the role played by the élite if we are to form a comprehensive picture of the democratization process.

From an élite perspective, it is a commonly held view that Chiang Ching-kuo's decision of 1986 to initiate political reforms was responsible for the KMT relaxing its grip on society and allowing democratization to take place. The process continued after Chiang died in early 1988, but the path to democracy was by no means smooth. In 1993, seven years after Chiang's decision, the constitutional set-up for the central government was still undecided, and the problem of confused national identity still cast a shadow over future political development. It is true that Taiwan can claim many significant achievements on the road to democracy, yet there are still some formidable barriers to be crossed. In particular, Taiwan seems to have encountered some unique problems that have caused the democratization process to drag on. Some insight into these problems might be gained from a study of élite attitudes and views towards the issues that have arisen during the course of political transformation, and the ability and willingness of the élite to resolve these issues. This chapter will identify the key issues facing the nation in the transition process and analyse the attitudes, stands, and actions of the political élite on all of them.

It is not an easy task to define the key political issues in Taiwan, since some issues which seem very important at a given time may quickly fade from view. But on the basis of élite interviews conducted over a number of years and the concerns of the academic community in Taiwan, it is possible to conclude that the key political issues are parliamentary reform, the constitutional setup for the new polity, and Taiwan's relations with mainland China. The issue of parliamentary reform may seem outdated since the mainland-elected parliamentarians were forced into retirement at the end of 1991. But the process of resolving it was dogged by controversy, and how members of the élite viewed the issue remains important in understanding the difficulties confronted by the nation in its quest for a fairer political system. A re-examination of the issue may throw some light on the question why Taiwan found it so difficult to resolve an issue that would have been deemed quite simple in other countries. The problem of constitutional revision, on the other hand, was an ongoing issue in 1993. Despite the revisions of 1991 and 1992, the constitution is still a topic of heated public debate. Finally, the most controversial issue and the most formidable barrier to democratization in Taiwan is its relations with mainland China, or, as it is generally termed in Taiwan, the national identity issue. This issue is at the heart of the debate between the two main political parties, and it has the potential to provoke an emotional confrontation in society.

Parliamentary Reform

Parliamentary reform was a central focus of debate in the course of political reform in Taiwan. The issue originated from the Nationalist government's retreat to Taiwan in 1949, when the ruling establishment, including the three chambers of the parliament, were transferred largely intact from the mainland.

When the term of the first parliament expired, the government amended the temporary constitutional provisions (emergency decree) to allow members to remain in office until the Chinese mainland was recovered. This 'frozen' parliament thus became a symbol of the KMT's claim to sovereignty over the Chinese mainland and its determination to return there one day. But as the mainland-elected members gradually died off and the average age of those that remained increased, the parliament became an important obstacle to democratization.

In 1969, the government held the first limited election for the three chambers of the parliament, allowing extra representation for the people of Taiwan to reflect the island's natural increase in population. Larger-scale supplementary elections were held from that time on, so that a small fraction of parliamentarians came up for re-election regularly, though the others remained in office in seeming perpetuity. The situation drew serious public criticism in the 1980s when the average age of the mainland-elected members was over eighty. In the late-1980s, the aging parliamentarians became such an embarrassment that the chambers of the parliament were removed from the itineraries of foreign VIPs.

The fact that almost all of the mainland-elected members were on the KMT side drew bitter criticism from the opposition, particularly after the lifting of martial law in July 1987. The DPP organized many public protests and created incidents within the parliament to publicize the issue. Although the extraordinary composition of the parliament provided the opposition with ammunition in its election campaigning, the existence of the senior members denied the opposition a chance of coming to power. Even if the opposition had won all the seats in a supplementary election, the KMT would still have been the majority party in the parliament. To put it simply, if Taiwan was to have any chance of becoming a democracy, the mainland-elected parliamentarians would have to be forced out.

The problem was, however, that while some government officials were eager to resolve the issue by any means, the KMT itself was facing a dilemma. Over the years, the mainland-elected parliamentarians had accumulated a great deal of leverage over national legislation because of the strategic position they occupied. For example, the emergency decree, which authorized the freezing of the parliament in the first place, could not be terminated without the approval of the National Assembly. And in another example, any government bill dealing with the question of pensions for retiring parliamentarians would need to go through the Legislative Yuan. Moreover, the government would have been in serious trouble if the senior legislators had decided to vote against other government bills. As expected, most mainland-elected parliamentarians strongly resisted all the forced retirement plans and refused to vote away their own political influence and financial privileges. The KMT had to tread a narrow path between public resentment on one side and resistance by senior parliamentarians on the other.

Faced with this dilemma, the KMT considered a plan to 'encourage' senior parliamentarians to retire by providing them with a substantial golden handshake. The plan was sent to the legislature in late-1988 in the form of the Voluntary Retirement Bill, and was formally adopted on 26 January 1989 (*The Journalist*, 99: 60). However, the Voluntary Retirement Act neither specified a deadline for retirement nor imposed any penalty for refusing to comply. Worse still, the act actually gave senior members an incentive to stay in their positions longer—as by doing so they would continue to receive their monthly stipends and their families would be able to collect their pensions after their deaths.

While some senior parliamentarians, tired of harassment from the opposition, wanted to retire as soon as the act was passed, the great majority insisted they would stay on. The *Central Daily News*, a newspaper owned and operated by the KMT, estimated at that time that the senior members would all be gone by 1999 (*Central Daily News*, 27 January 1989: 1). This was based on the premise that they would agree to retire within the ten-year period. This was rather optimistic, however, as interviews conducted by the same newspaper suggested that at least one hundred of the senior parliamentarians 'would not retire under any circumstances' (*Central Daily News*, 29 January, 1989: 2; 31 January 1989: 2). These refusals caused the government further embarrassment, though they shoved aside the opposition argument that there must be a penalty clause attached to the act and stubbornly insisted on giving the parliamentarians strong financial incentives not to retire. As the issue of parliamentary reform lingered on it was apparent that more drastic measures were needed.

The Stands

The freezing in office of a large proportion of the parliament was a focus of conflict between the KMT and the DPP, since most of these senior members were affiliated to the ruling party. Dramatic events, such as pushing and shoving on the floor of the Legislative Yuan, got frequent press coverage at home and even occasionally abroad. Some of the largest political demonstrations in Taiwan, such as those held in June and December 1987, and March 1988, were specifically aimed at the issue of parliamentary reform, and they won substantial public support. As was revealed in personal interviews, the opposition élite's stand on the issue was quite straightforward: all senior members of the parliament must resign and the government must hold a comprehensive general election (*quanmian gaixuan*). Among the KMT élite, however, there was no consensus as to how the issue should be resolved.

The DPP

For opposition lawmakers who had to face the senior legislators on a daily basis, their presence in the Legislative Yuan was the number one problem to tackle. The pressing need to reform the parliament was perhaps best stated by You Ching, a former member of the Control Yuan and later the Legislative Yuan who was elected magistrate of Taipei County in 1989 and again in 1993:

> Right now our biggest problem is the overall re-election of the parliament. Without overall re-election, democracy is just impossible. . . . If the KMT is sincere in making changes, it should hold an overall re-election at least for the Legislative Yuan. Of all the issues in Taiwan's democratic transition, the issue of a general election is the most important. . . .
>
> Right now the KMT is making all kinds of excuses for these old parliamentarians, such as the authorization of the emergency decree or the decision of the Council of Grand Justices. But these are all nonsense. It is clearly stated in the constitution that the Legislative Yuan should be re-elected every three years and the National Assembly and Control Yuan every six years. There is no excuse for not holding general elections. (personal interview, 26 August 1988)

You's view on the issue was typical of DPP leaders at the time. And the response of opposition leaders in personal interviews was generally in line with public sentiment on the issue. In a public opinion survey conducted by the *Independence Evening Post* in April 1987, 51.4 per cent of respondents saw a general election as the solution to the problem.[1] If respondents born in mainland China were excluded, those in favour of a general election rose to 54.1 per cent, with only 28.9 per cent against (*Independence Evening Post*, 14 April 1987: 2).

However, the views of the DPP élite or the public counted for very little as the KMT was the only political force with sufficient power to resolve the issue. All the DPP could do was rally public support to bring pressure on the government. Some opposition leaders despaired of the KMT ever taking any action and they drew comfort from the knowledge that nature would sooner or later take its toll and the senior parliamentarians, whose average age was eighty-three when the Voluntary Retirement Act was adopted in 1989, would die out. The question was, however, whether the public's patience would last that long.

The KMT

In contrast to the opposition élite who unanimously demanded a general election, the KMT side showed a high degree of divisiveness on the issue. The more liberal members of the KMT élite quietly agreed with the DPP's general election proposal. Lin Yu-hsiang, a KMT legislator and the leader of the Wisdom Coalition, was one of the very few members of the KMT élite willing to stand up and demand the retirement of the senior parliamentarians. However, his open criticism of the party was too much for the old guard, and Lin was called in front of the party's disciplinary committee, which was dominated by conservatives. When the committee deprived him of all rights within the party for two years, Lin retorted that the party had sacrificed him to please the 'old thieves' (*laobiao* or *laozei*) in the parliament.[2]

Disciplinary action did not deter Lin Yu-hsiang from criticizing the party. On the contrary, he was greatly encouraged by the popular support he attracted for his attacks against the senior parliamentarians. He went so far as to argue for a general election during the 1989 election campaign and made it his number one campaign promise to bring down the senior parliamentarians. Few would doubt that his DPP-style campaign against the KMT helped him get re-elected in 1989. Lin was not alone in the KMT. All the members of the Wisdom Coalition sharply criticized the existence of the senior parliamentarians during the election campaign and blamed the KMT for not resolving the issue. Their stand revealed that younger Taiwanese elected representatives in the KMT were reform-oriented and willing to distance themselves from the party establishment. It also showed that there was a group of KMT lawmakers who might enter into an alliance with the DPP to push for parliamentary reform.

Government officials in Taiwan usually exercise more restraint when it comes to voicing opinions on controversial issues, for any open statement considered to be against KMT policy might endanger their promotion prospects. However, public restraint did not necessarily imply agreement with official policy. Official KMT policy on the issue during the period when the Voluntary Retirement Act was adopted was that senior parliamentarians should be permitted to retire gradually and honourably, and that they should not be harassed in any way. During the author's interviews with government officials, some tried to interpret the KMT's official stand, but others, particularly younger officials, revealed their personal disagreement with the policy and insisted that the mainland-elected parliamentarians must be forced into retirement. One KMT legislator and two government officials went so far as to argue for a general election, which went directly against official policy at the time.

Lee Teng-hui, who was responsible for adopting the policy of voluntary retirement, and Ma Ying-jeou, the architect of the Voluntary Retirement Act and the KMT official chiefly responsible for convincing the senior parliamentarians to accept the retirement plan, were quite optimistic about the voluntary measure, and they were convinced that the parliament could be completely reformed without a 'revolution'.

Yet in the same interviews, both Lee and Ma showed their understanding of public feeling on the issue and accepted that the parliament must be reformed if the goal of democracy was to be attained. Their advocacy of 'voluntary' retirement was apparently a compromise with the conservatives who were fighting bitterly against the pressure for change. To the conservatives, who were still in important positions in the party and government institutions, the mainland-elected representatives were the symbol and the source of the government's legitimacy, both in ruling Taiwan and claiming sovereignty over mainland China. They were the final link with the mainland and without them Taiwan would be as good as independent. Thus the conservatives were implying that if the government pushed for parliamentary reform, it would be pursuing a policy of permanent separation from China, a position still deemed unacceptable and unforgivable by the KMT as a whole, and therefore to be carefully avoided by any policy-maker.

The more liberal members of the KMT élite fully acknowledged the position held by the conservatives. But their response was that the senior parliamentarians had been elected so many years ago that they represented neither the people of Taiwan nor the people of the mainland. They further argued that for Taiwan to become an attractive model of democracy to mainland China, the government should pursue thorough democratization. After all, as one KMT official said at the time, 'the senior parliamentarians will all be dead in a few years. Are we really going to lose our legitimacy when that happens?' (personal interview, 2 August 1989).

Resolving the Issue

As generally predicted by the public, the voluntary retirement plan was not well received by the senior parliamentarians, and only a few retired voluntarily, some of them filing their retirement applications on their death-beds. The others continued to receive their monthly stipends in the face of public criticism. It was obvious that money was not a strong enough incentive for retirement.

As the presidential election of early 1990 approached, it was clear that the members of the National Assembly, the body designated by the constitution to elect the president and vice-president, were still determined not to budge. Indeed, large numbers of them took the opportunity to strengthen their position against the government reform plan and aligned with the group in the KMT challenging Lee Teng-hui's leadership and criticizing the president for pursuing Taiwan independence in substance (*dutai*) and dictatorship (*ducai*). As mentioned earlier, the crisis caused a public uproar and led to a week-long student protest and street violence by the most radical elements of the opposition.

Lee, however, managed to get elected, and in a very skillful maneuver turned the mounting public pressure for reform to his own advantage. In accordance with the students' demands, he called a National Affairs Conference in June 1990 to discuss the issues of parliamentary reform, revision of the constitution and the composition of the central government, and relations with mainland China. Even though the conference was criticized by many as no more than a debating forum whose decisions were not binding, it did provide Lee with a mandate to rescind the emergency decree on which the legitimacy of the senior parliamentarians was based.[3]

Unlike the discussions on other issues which were quite controversial, the debate in the National Affairs Conference on parliamentary reform focused on when, rather than whether, the mainland representatives should go. But in order to assuage the conservatives, the KMT side proposed that the National Assembly and Legislative Yuan should include a number of at-large seats (*quanguo bufenqu daibiao*) to be distributed among the parties according to their proportion of the total vote. This would enable the KMT to ensure that the parliament contained a certain proportion of mainlanders, thus continuing the legitimacy of the government's claim to sovereignty over the mainland. The old National Assembly met in April 1991 to pass the additional articles of the constitution which would replace the emergency decree. Thus the KMT used the senior assemblymen for one last time to rescind the emergency decree and end the term of the senior parliamentarians. The additional articles were promulgated by the president on 1 May 1991, and the emergency decree (Temporary Constitutional Provisions for the Period of General Mobilization Against Communist Rebellion) was rescinded simultaneously. At the same time, the Council of Grand Justices resolved the parliamentary reform issue by ruling that the senior parliamentarians should all resign by the end of 1991.

The ruling of the grand justices was a significant blow to those whose special privileges were tied to the unreformed system. The conservatives in the KMT, represented by Legislative Yuan speaker Liang Su-yung, claimed that the ruling was unconstitutional. Liang, who declared that he for one would not step down, became one of Lee Teng-hui's most vocal opponents, by frequently accusing him of promoting Taiwan independence in substance and trying to institute a dictatorship.

Thus the rift between President Lee and his KMT challengers was seen by the public as a conflict between reformers and conservatives. Yet the resistance of the senior parliamentarians became irrelevant after the emergency decree was rescinded. Resolving the issue of parliamentary reform was indeed a significant step forward in Taiwan's democratization.

Constitutional Reform

A constitution, the codification of government institutions and basic rules for government operation, is one of the most important foundations of a democratic political system. A constitution is also a legal document of the highest authority in a democracy, and any other law or government directive is nullified if it is found to contradict the constitution. The existence of a constitution is not a sufficient condition for the successful functioning of a democratic political system, but it is generally recognized that democracy will not succeed without a set of commonly accepted rules to regulate political processes and to map out political power.

Writing a new constitution or revising an existing one is of special importance in an authoritarian country undergoing democratization. In order for the political forces in the new and democratic government to compete with one another fairly and for a government to be formed without conflict and bloodshed, a new constitution is necessary. Yet precisely because the establishment of a new political system involves the restructuring of political power and national resources, the process of writing or revising a constitution is almost always accompanied by intense debate and controversy. Any group with a vested interest in the authoritarian system is very likely to stand in the way of change, while those who were deprived of power will use all means to guarantee that their interests are fairly represented in the new system. Constitutional revision in Taiwan has taken place against this kind of background, but its process is very different from that of other democratizing countries.

The ROC constitution, written in 1947 in China, prescribed a central government structure seen by the public in Taiwan as hampering the process of democratization.[4] This central government system is neither presidential like that of the United States nor parliamentary like that of Great Britain, nor is it the mixed system found in France. In Taiwan, neither the president nor the premier is chosen by the public in open popular elections. Yet the president is the commander-in-chief of the military, has the power to declare a national emergency, and most important of all, must co-sign all laws and orders issued by the premier (Sections 35 to 43). The constitution also specifies that when there is a conflict between the five Yuan, the president has the power to mediate so that the government can continue to function (Section 44). The constitutional provisions governing the presidency, therefore, provide the president with an enormous amount of power over government decision-making.

At the same time, the premier is not merely the president's deputy. The ROC constitution clearly specifies that the Executive Yuan is the highest administrative organ of the nation, with the premier as the chief executive (Section 53). The Executive Yuan has the power to propose legislative bills and budget bills, treaties, and declarations of war and martial law to the Legislative Yuan (Sections 57, 58). According to the constitution, the Executive Yuan should report to the Legislative Yuan on important government decisions and policies, and the members of the Legislative Yuan have the right to interpellate ministers, including the premier. Moreover, if the Legislative Yuan is dissatisfied with a particular bill, it can send it back to the Executive Yuan for revision, though the Executive Yuan may, with the approval of the president, refuse to rewrite the bill. In a similar way, the Executive Yuan may refuse to accept revisions made by the legislature to one of its bills. In both cases, however, if a two-thirds majority of legislators supports the Yuan's original decision, the premier should either accept it or resign (Section 57–2). This provision is clearly similar to the no-confidence vote found under a parliamentary system. Yet one important factor distinguishes the ROC's five-power system from the parliamentary model: the premier and the ministers are not members of the Legislative Yuan. Rather, the president appoints the premier who in turn appoints the ministers subject to presidential approval. While the president is indirectly elected through the National Assembly, the premier is not held accountable to any public selection process, except that he has to be confirmed by the Legislative

Yuan. In short, this is a very strange set-up for a central government and is the most important area that needs to be reformed if Taiwan is to make further progress with democratization. For Taiwan to be a true democracy, at least one of the two top executive offices *must* be open to a popular electoral contest.

Another area where political problems have arisen is that of the National Assembly and the Control Yuan. Although these two institutions are very different in appearance and perform very different functions, they faced similar criticism from the general public. The National Assembly, which until 1992 was packed with mainland-elected deputies, met only once every six years to elect the president and vice-president, yet its members assumed all the privileges of members of parliament. As the student participants of the March 1990 protests argued, the people of Taiwan were mature enough to elect their own president and did not have to waste national resources to retain the National Assembly for that purpose.

The Control Yuan's function as an anticorruption watchdog was seriously undermined by the fact that the KMT monopolized the selection of its deputies. Moreover, the endless corruption allegations against the Yuan's Taiwan-elected deputies after 1969 aroused public distrust and even distaste for the body. The KMT argued, however, that the National Assembly and the Control Yuan were inseparable parts of Sun Yat-sen's five-power constitutional design and for that reason they must be preserved.

Resolving the Issue: The First Attempt

Several panels of the National Affairs Conference were devoted to discussion of the central government system: whether it should be based on a presidential or a parliamentary model and how the president should be elected, and whether the National Assembly and the Control Yuan should be abolished (see Table 5.1). As mentioned in the previous chapter, this conference was a forum in which different groups and sectors of society could present their views; its decisions were not binding on the government. On the issue of the National Assembly alone, some Conference participants argued that the body should be abolished, while others (most of whom were

Table 5.1 Issues and Stands in the National Affairs
Conference, June 1990

Issue	Stand	No. of Supporters
National Assembly	Abolish National Assembly	44
	Turn it into electoral college	65
	No change	15
Method of Electing President	Direct popular election	52
	Through electoral college	65
	No change	20
Central Government System	Presidential	36
	Cabinet	7
	Dual leadership	57
	No change	28
Control Yuan	Abolish Control Yuan	38
	Quasi-judicial institution	37
	No change	34
	Turn it into upper house of parliament	4
Examination Yuan	Abolish Examination Yuan	37
	Process national examination only	17
	No change	42

Source: *Independence Evening Post*, 3 July 1990: 2.

on the KMT side) argued that it should be retained, though
its functions should be revised to make it similar to the US
electoral college. As for the Control Yuan, the opposition-
oriented delegates to the Conference dismissed it as totally
unnecessary in a democracy and recommended that it be abolished. Yet delegates who sided with the KMT considered that
it was essential to retain the five-power constitutional framework, and argued that the Control Yuan should be made into
a quasi-judicial body, appointed rather than elected, which
would oversee the ethics of government officials.

On the issue of whether Taiwan should adopt a presidential

or a parliamentary system debate was also quite intense, and it was focused on the method of electing the president. Many argued for direct popular election, but the KMT participants hesitated to go along with this as a popular election would deprive the National Assembly of its prime function and therefore make its abolition inevitable. This particular panel was chaired by Frank Wu, the director of the *Independence Evening Post* and an opposition sympathizer, and he concluded the session by declaring that the president should be directly elected by the people, which evoked protests from the KMT participants.

The National Affairs Conference was the first occasion in Taiwan's recent history that the head of state had called a large-scale meeting to discuss critical issues facing the nation. But because of its size (there were more than one hundred local and overseas Chinese delegates) and the lack of consideration given to the need to balance the various political forces, the conference became no more than a forum for debate. Public hopes that the conference would pave the way for revision of the constitution were destroyed when it was announced that the views presented would only be referred to by the president in carrying out further political reforms. It was clear that other steps would have to be taken before Taiwan's political institutions could be made democratic in essence.

Resolving the Issue: The Second Attempt

An opportunity for further reform came in May 1991 when the National Assembly passed the first constitutional amendments and rescinded the emergency decree, and the Council of Grand Justices ruled that the mainland-elected members of the parliament must step down by the end of the year. These steps cleared the way for the election of a new National Assembly with the specific task of revising the constitution.

The election of the new 325-member National Assembly took place at the end of 1991 as scheduled. The KMT won 254 seats, the opposition DPP 66 seats, and the independents 5 seats (*China Times*, 22 December 1991: 1). But the situation was complicated by the fact that the 78 deputies who won their seats in the 1986 supplementary election were also included in the second assembly until their terms expired in

1992 (*United Daily News*, 23 April 1990: 2). That put the total number of deputies at 403. This number was exceptionally large both relative to Taiwan's population and compared to parliamentary bodies elsewhere, making it difficult for the Assembly to reach any concrete decisions.

The KMT claimed that the results of the 1991 election gave it a clear mandate to direct the constitutional revision process (the party had gained 71 per cent of the popular vote and more than 78 per cent of the total seats in the assembly). This allowed the KMT to dominate the assembly sessions, since the constitution provided that a three-fourths majority was necessary to pass a constitutional reform proposal. But it was obvious that this would only hold true if the KMT members of the National Assembly remained as a united group and were willing to follow the orders of the party centre, and if the party centre in turn had a unified agenda for constitutional revision. However, KMT solidarity was put to a serious test when the National Assembly convened in March 1992.

Even though the legitimacy of the new National Assembly was recognized by the general public on account of it having been elected democratically, its exceptionally large size ensured that its sessions were chaotic. Frequent disruptions caused by procedural problems, shouting matches, and pushing and shoving prevented the assembly members from addressing key constitutional issues.[5] The tense atmosphere between the parties, and even between factions of the same party, did not allow issues and stands to be clearly represented.

The chaos in the National Assembly could have been significantly alleviated had the ruling party itself reached a binding consensus on key issues such as presidential elections and the selection of members of the Control Yuan. Group theory would have us believe that 'the greater effectiveness of relatively small groups—the "privileged" and "intermediate" groups—is evident from observation and experience as well as from theory' (Olson 1977: 53). This is also emphasized in studies of the Spanish experience of constitution-making:

> Party élites . . . must have the capacity to obligate their followers to respect or support that compromise before it may be regarded as successfully resolving the conflict in question. . . . In the parliamentary context, this requires the existence of party discipline. Outside of parliament, it requires that a party

have the capacity to control the behavior of its supporters.
(Gunther and Blough, 1981: 395)

During the National Assembly election campaign, the KMT
had made it quite clear that the president should be elected
indirectly through the National Assembly. This stand was in
very sharp contrast to that of the DPP, which argued all along
that Taiwan should adopt a presidential system like that of
the United States, and that the president should be popularly
elected. But a few weeks before the National Assembly convened in March 1992, the issue of the presidential election
suddenly became a topic of heated debate within the KMT's
CSC, as if its previous position had been an electoral ploy
rather than a serious campaign promise or a unified party
stand on the issue. Because of the KMT's absolute majority,
the new National Assembly became the battleground for an
intensified power struggle between the KMT factions that had
formed during the 1990 presidential election campaign, and
the conflict with the DPP was set aside. This is an obvious
example of how the ROC's experience of constitution-making
differs from that of other countries.

In view of its minority position, the DPP decided that its
strategy in the National Assembly should be—in Albert O.
Hirschman's (1970) words—to 'voice' its discontent at the
KMT's domination of the process and to threaten to 'exit'
the system. Even though the DPP's long-standing position on
the separation of powers between the three branches of government and the direct election of the president was clear
and sound, as opposed to the KMT's unclear position on constitution revision and unwillingness to depart sharply from
the problematic five-power system, its minority position prevented it from making a significant impact on the drafting of
the new constitutional provisions. It could only engage in
filibustering and raise procedural issues to attract public attention to its grievances. The opposition party also held rallies
and demonstrations, those of 19–24 April being the most
noticeable, to voice its discontent over the constitutional revision process. At times it also threatened to boycott the process,
a quasi-exit option (Hirschman, 1970: 86), to make it clear
that the KMT alone should be responsible for the constitutional mess.

In early May, the DPP deputies finally withdrew from the

National Assembly. Three weeks later, eight members of the Independents' Alliance also dropped out, leaving the constitutional revision process entirely in the hands of the KMT. The withdrawal of the opposition was understandable on account of its inability to influence the process. The DPP originally had high hopes of forming a temporary coalition with some KMT members to push for the direct election of the president. But this was forestalled when the ruling party, because of dissent within its ranks, decided to shelve this issue until 1995. Unable to influence the outcome, the opposition and its independent allies had no incentive to remain in the National Assembly and share the blame for not reaching a clear decision on this issue.

The opposition's withdrawal did not end the conflict in the National Assembly or improve the general atmosphere of the session. The KMT deputies were engaged in a tug-of-war on various issues, particularly on the proposal to increase the power of the Assembly and reduce that of the Legislative Yuan. The resulting chaos led to calls from the public for the abolition of the National Assembly altogether.

The Stands

Since the process of democratization was set in motion, there has been evidence of a growing conflict of political interests. Because the KMT is in the process of breaking away from its authoritarian past, many of its legislators and other elected representatives are eager to express their own ideas on the decision-making process and challenge the authority of the CSC, most of whose members are not elected representatives. In other words, the weakening of the party centre, a result of the changing political environment, has coincided with the growing autonomy of the representatives, including legislators and members of the National Assembly, who correctly claim to represent a certain proportion of the population.

Moreover, the electoral system and the nomination process also contribute to the growing factionalism within the ruling party. For the party to maximize its gains in a multi-member district, the KMT often has to form coalitions of convenience with local factions and nominate candidates who have significant influence in local politics. But the interests of the local factions may not always correspond with the interests of the ruling party at the national level. As a result, the elected KMT parliamentarians, who have spent a tremendous amount of money and gone through great difficulties to win their seats, may well place their personal and factional interests ahead of the party's interests. They also form factions within the parliament to make sure their interests prevail. During the constitutional revision process, the party centre might have preferred direct presidential elections and wished to limit the power of the National Assembly. However, it would have been inconceivable for the popularly elected members of the National Assembly to support the party's stands on these issues, as to do so would have meant committing institutional suicide.

The KMT

Because of the very pluralistic nature of Taiwan society and its conflicting political interests, it is difficult to divide the significant actors in the political arena into all-inclusive and mutually exclusive categories. As outlined in Chapter 4, both of Taiwan's major political parties are plagued with factional wrangling. The KMT split into Mainstream and Non-mainstream factions in 1990, and since then the factional squabbles have continued and even spread to other political institutions. The pro-Mainstream Wisdom Coalition published its own ideas on constitutional revision as early as April 1991. The coalition advocated direct election of the president and the abolition of the National Assembly and the Control Yuan, making the Legislative Yuan the only chamber of the ROC's parliament.

Before they broke from the KMT to form the New Party in August 1993, the legislators of the New KMT Alliance were staunch supporters of the more conservative Non-mainstream faction, which also had the support of a dozen or so legislators with military backgrounds or military backing. The New KMT Alliance strongly opposed President Lee Teng-hui both as president and chairman of the KMT. However, prior to August 1993 when the New Party announced its support for a presidential system and a proper checks-and-balance mechanism for the central government, the Alliance could not decide whether the president should be directly elected and the National Assembly abolished, or whether the five-power system should be retained. The Alliance revealed its distaste for the National Assembly, however, when its leader, Legislator Lee Sen-fong, called Wang Ying-chieh, a leading KMT member of the Assembly, a 'piece of trash' (*China Times*, 14 April 1992: 2). The subsequent attacks and rebuttals clearly showed that the New KMT Alliance did not want the government to retain the National Assembly. There were other groupings in the Legislative Yuan, such as the Construction Research Association (*Jianshe yanjiu hui*) and the Association for Reform of the Parliament (*Guohui gongneng gaige hui*), but they were loose organizations without much political significance.

In the National Assembly, the KMT was divided by the issue of presidential election, as the names of the two factions, the Direct Election faction (*Zhixuan pai*) and the Indirect Election faction (*Weixuan pai*) indicate. The two factions had emotional confrontations on the floor of the Assembly over how the president should be elected and whether political power should be centered on the president or the premier, and they also tried to undercut each other's power. However, on other issues, the two factions were not so clearly divided. Many members of both factions worked hard to increase the power of the National Assembly by proposing that it should institute a speaker (*yizhang*), the equivalent of the majority leader in the US House of Representatives, and committees and subcommittees, which would make it a permanent institution instead of an electoral college on the American model. They also wanted the Assembly to have the power to review legislation and confirm presidential nominations to the Control Yuan and the Council of Grand Justices. What made the situation worse was that many National Assembly deputies wanted to restrict the power of the Legislative Yuan by overseeing its budget and shortening its term of office from three years to two. This latter proposal evoked violent protests from the legislature which threatened to cut the National Assembly's budget. Consequently, the KMT's original proposal to extend legislators' terms to four years had to be shelved because of opposition from the National Assembly.

The DPP

On the DPP side, the traditional conflict is between the more moderate Formosa faction and the radical New Tide. These two factions also differed on the issue of constitutional reform,

Formosa stressing the need for direct election of the president, and New Tide demanding a new constitution, a new political system based on the American model, and a new name for the country: the Republic of Taiwan.

On the eve of the 1991 National Assembly election, the DPP issued a 'Draft Constitution for Taiwan' (*Taiwan xianfa caoan*). The system envisaged by the opposition and set out in this draft constitution and the 'Magna Carta for Democracy' (*Minzhu da xianzhang*), issued in 1990 in preparation for the National Affairs Conference, was a presidential system with a single-chamber parliament and a mechanism of checks and balances between the executive and the legislature. Debate at the party's congress in 1991, however, focused on the title of the draft constitution and the name of the country. Members of New Tide wanted to call the document a 'Draft Constitution for the Republic of Taiwan', and they insisted that the name of the country be stated clearly in section one. But the Formosa leaders thought that the New Tide proposal was too radical to gain public acceptance and feared that it would harm the party's performance in the upcoming election.

New Tide won the ensuing debate and the DPP was forced to make independence from mainland China the key issue in its 1991 election platform. Moreover, the two factions seemed to be able to contain their differences during the constitutional revision process in 1992 and coordinated their activities both in the National Assembly and on the streets.

The Issue Lingers

With great discrepancies between parties, institutions, and factions, and with each trying to play a role in the constitution revision process, it was very difficult to reach a consensus. Neither of the two political parties, particularly the ruling KMT, were able to work out their internal differences and reach decisions binding on their members in the National Assembly. Even a temporary coalition between the KMT's Mainstream faction and the DPP's Formosa faction would not have given supporters of direct presidential election the necessary three-fourths majority to get their proposal through the Assembly. Moreover, for a KMT faction to ally with the DPP on that issue, a compromise would have had to be reached with another KMT faction on some other issue. The result of the compromise might not have been what the coalition wanted. In addition, such coalition efforts would very likely have antagonized other party members and caused the parties to split. Splits would have made the situation even more complicated and a consensus even harder to reach. This kind of political rivalry eventually made it impossible to revise the constitution to suit Taiwan's needs.

The KMT's final plan for constitutional reform was approved by the CSC on 6 May 1992, and sent to the National Assembly for adoption (*China Times*, 7 May 1992: 2). The most important feature of the nine new provisions was that they significantly increased the power of the president and the National Assembly at the expense of the premier and the Control Yuan. The president, under the new design, has the power to nominate members of the Control Yuan, the Judicial Yuan (including the Council of Grand Justices), and the Examination Yuan, and his nominations are subject to confirmation by the National Assembly. However, this reallocation of power lacks certain supporting features and makes the new constitutional setup look very strange.

First of all, the key issue of the method of electing the president was left undecided, leaving room for further factional conflict and another political crisis. If it is decided at a later date that the president should be directly elected, there will have to be another round of revisions to delete those articles in the original constitution relating to presidential election. This would make the National Assembly virtually redundant. Despite this possibility, the KMT's revision proposals give the Assembly new powers to confirm certain presidential nominations. A fair explanation for this apparent contradiction is that the KMT centre wanted to give the National Assembly additional powers to make up for possible future losses in other areas, and by doing so ensure the smooth adoption of its nine provisions.

Secondly, the distribution of power between the president and the premier, which sparked off the constitutional crisis of 1990–91, was not even touched upon in the KMT provisions. The DPP and the non-partisan National Assembly members had high expectations that the reform process would produce a presidential system of government with the premier as the president's chief administrator. However, this idea was blocked by KMT members of the National Assembly, prompting the walk-out of DPP members and independents. The KMT, left to design its own revisions, had to shelve this issue on account of bitter factional rivalry. The new provisions thus failed to resolve the most important issue that had given rise to the original constitutional crisis.

Another strange feature of the provisions is the method for selecting members of the Control Yuan. The Control Yuan had come under fierce public criticism for corruption, and the KMT's CSC had decided to deal with this by having Yuan deputies appointed by the president instead of elected either by the Provincial Assembly or municipal councils or by the entire electorate. However, this change brings into question the Yuan's integrity and its ability to oversee the operations of the government of which the president is part. In addition, one may reasonably suspect that the president will choose Control Yuan deputies on the basis of loyalty rather than merit or party balance.

In short, the great efforts put into the constitutional revision process did not resolve the problems stemming from the old constitution and thus prevent a future constitutional crisis; the whole process may therefore be judged a failure. Moreover, it added new uncertainty to an already complicated system. To establish a central government system that is democratic in essence will take another round of conflict among political parties and debate among the élite.

Independence versus Unification

To many members of the national political élite in Taiwan, the most important yet divisive issue underlying democratization is conflicting interpretations of Taiwan's national identity. People in Taiwan have very different views regarding whether Taiwan is a part of China (and its government is the legitimate ruling power of the entire country) or whether it is a separate political entity. This difference is the cause of many of the difficulties involved in parliamentary reform, revision of the constitution, the termination of the emergency decree, the direct election of municipal mayors and the provincial governor, and the KMT's continuing domination of political resources. Without the problem of national identity, all these other issues might have been resolved long ago.

At one end of the spectrum of debate, those who believe Taiwan is part of China and that the government in Taipei represents the Chinese people on the mainland have argued that Taiwan should maintain representatives from mainland China in its parliament, should not hold direct presidential elections, and most important of all, should not amend the constitution or terminate the emergency decree (*Central Daily News*, 24 April 1989: 2). They have argued this on the premise that Taiwan and China will be reunited some day and the Nationalist government will be the legitimate ruler of China. Although Taiwan has been separated from China since 1949 (and also during the Japanese colonial period between 1896 and 1945), and the chances for unification on Taiwan's terms in the near future are slim, the Nationalist government maintains its claim to sovereignty over the mainland. It was a perceived need to preserve the legitimacy of this claim that made the Nationalist government hesitate to dissolve the mainland-elected parliament or completely rewrite the constitution.

At the other extreme, those who think of Taiwan as an independent political entity view the political set-up during the four decades after 1949, including the parliament, the governorship, and the constitution, as the product of a twisted interpretation of Taiwan's international status. The only purpose of this interpretation, according to the opposition élite, is to bolster an undemocratic political establishment biased towards the KMT.

In dramatic contrast to the experience of most Third World countries which have struggled for independence from colonial powers, the élite in Taiwan are divided over whether the country should unite with another much larger, more powerful, but much poorer country. Without a consensus on Taiwan's national identity, the debate on democratization focuses on nationalism, not the specific structural issues that need to be resolved.

The debate over national identity is one of the most divisive issues facing Taiwan today, and the emotion attached to the issue clearly threatens to polarize society. To many people, the fundamental problem of Taiwan's political system is this unrealistic claim to sovereignty over the mainland, as without such a claim the KMT would have no justification for maintaining the existing central government set-up. One solution, favoured by many, is that Taiwan should declare itself independent from China. However, to many others, unification with the mainland is the last hope, and independence is an unacceptable alternative as it would mean national betrayal. An element of fear is added to this debate by the repeated threats from Beijing that the PRC will invade if Taiwan ever declares itself independent.

The question 'Do you prefer unification between Taiwan and China or an independent Taiwan?' was so sensitive prior to Chiang Ching-kuo's reform announcement in 1986 that no public opinion survey in Taiwan had attempted to find an answer. Since the beginning of the reform period, however, several versions of the independence/unification question have been posed by various survey organizations, and from the results it is possible to see how divided the public are on the issue. A survey conducted by the short-lived *Capital Morning Post*, published by opposition legislator Kang Ning-hsiang, was probably the first to ask a direct question on this issue. About 12 per cent of respondents wanted Taiwan to be independent of China, and of those, about 40 per cent wanted independence regardless of whether it would provoke an attack from the mainland. Only about 5.4 per cent of respondents wanted the Taiwan government to unify China,[6] and another 56.3 per cent wanted to maintain the status quo, which meant a continuing stalemate between Taiwan and China. Moreover, about 90 per cent of respondents did not want Taiwan to be ruled by the Chinese communists (*Capital Morning Post*, 29 July 1989: 2).

In another survey conducted by *The Journalist* in September 1988, respondents were asked, 'When you go to mainland China, do you feel you are going back to your motherland or going to a different country?' To this question, 39.4 per cent of respondents answered 'motherland', and about 30.5 per cent replied 'a different country'. If provincial origin is held constant, about one half of the mainlanders thought of China as their motherland, as opposed to about 36 per cent of Taiwanese (*The Journalist*, 81: 41–2).

A survey conducted by *Global Views Monthly* (*Yuanjian*), a news magazine supported by the KMT, found that half of all respondents considered themselves 'Taiwanese', as opposed to 35 per cent who considered themselves 'Chinese' (*Global*

Views Monthly, 1 July 1987: 35). In the same survey, only about 32 per cent of the Taiwanese thought of themselves as 'Chinese', and only 18 per cent of mainlanders considered themselves 'Taiwanese'. Another survey conducted in 1987 by an independent poll organization asked respondents whether they supported the idea of an independent Taiwan. It is interesting that almost 10 per cent of respondents replied that they supported the idea, even though advocating an independent Taiwan was still deemed a criminal offense at that time (*China Times*, 7 August 1987: 2).

The independent Public Opinion Research Foundation conducted a series of surveys on the issue of Taiwan independence in 1989, at the height of the debate and shortly before the Legislative Yuan election. When asked whether people in Taiwan were free to express the idea of Taiwan independence, 32.3 per cent of respondents said 'yes' and 33.2 per cent said 'no' on 6 November, while 38 per cent said 'yes' and 31.4 per cent said 'no' on 7 November. When asked on 10 December whether election candidates should be allowed to promote Taiwan independence in public, 17.1 per cent of respondents answered 'yes', while 48.5 per cent answered 'no'. On 14 May, the question 'Do you agree with the idea of Taiwan independence?' received a positive response from only 2.7 per cent of respondents in military dependents' quarters (*juancun*), whereas 91.3 per cent said 'no'. Among the general public, 15.8 said 'yes' and 57.3 said 'no' on 3 September, and the percentages were 9.7 per cent and 59 per cent on 6 November, 15.7 per cent and 59.5 per cent on 7 November, and 9.2 per cent and 62 per cent on 10 December.[7]

The results of these public opinion surveys, which show that there is a certain segment of society that supports Taiwan independence but another large segment which is opposed to the idea, indicate that Taiwan society is indeed facing an identity crisis over Taiwan's peculiar relations with China. There is no consensus on whether Taiwan is a part of China, or whether Taiwan should unify with China. Not many people support a declaration of independence, yet people who identify with China as their motherland are not in the majority either. Furthermore, no one is sure how many of those who think of China as their motherland really support the idea of unification. Perhaps most people in Taiwan want neither of the two goals, both of which threaten to ruin Taiwan's prosperity and relative freedom. The debate among the political élite over the issue can be viewed as an extension of public differences over Taiwan's national identity. Members of the élite may be better informed on the issue than the public, but they are just as divided on it.

Moreover, the issue of national identity cannot be resolved entirely by the Taiwanese themselves. The periodic threats by the PRC to invade if Taiwan declares independence have evoked different responses in Taiwan. Some members of the opposition élite think that independence is the only way to overcome the various obstacles to democratization, and further democratization in Taiwan should not depend on China's attitude. They argue that the KMT's previous policy of retaking the mainland did not provoke an attack from Beijing, so why should independence, which in essence means giving up the claim to sovereignty, have that effect. The Taiwan military, they charge, claims to be in a state of readiness for a Chinese invasion, therefore an attack on Taiwan would inflict a heavy human and economic toll on China itself. They believe that a Chinese attack is particularly unlikely at a time when China is trying hard to modernize its economy and society and has few spare resources for a military adventure. Ardent supporters of Taiwan independence also argue that a declaration of independence would raise the issue of Chinese military action to an international level—that of one sovereign state committing an act of aggression against another—and this would be the only way that Taiwan could call on international support in the event of a Chinese attack.

However, the more pragmatic opposition leaders and the KMT élite consider the above argument to be dangerously irresponsible. Risking war with a military giant only to achieve *de jure* independence, they say, is unnecessary, since Taiwan's *de facto* independence, its political system and economic development, can be better protected by not provoking China. And it is quite true that one should not experiment in politics, especially when a policy might provoke fierce internal resistance and foreign aggression. If China were to begin any military action, be it a large-scale attack or just coastal harassment, Taiwan's trade- and finance-oriented economy would soon be in shambles.

The élite response to unification is also quite drastically divided. Since both the KMT and the CCP have made unification their official policy, the opposition élite and some more liberal KMT leaders claim that the KMT is playing into Beijing's hands. No matter how many unification models the KMT proposes, they argue, as long as it claims that there is only 'one China', it is the PRC, not the ROC, that will be recognized by the world as representing the Chinese people. They claim that the Taipei government's insistence on its claim to Chinese sovereignty has also made it difficult for the international community to recognize both China and Taiwan and allow Taiwan to have the same status as China. Even though in the 1990s the Taiwanese government has engaged in a pragmatic and flexible diplomatic policy and made it known that it is willing to coexist with Beijing, Beijing has not responded in kind by allowing other countries to maintain dual recognition, and this attitude is not likely to change in the future. This section of the élite argues that as a result, Taiwan's international activities are held hostage by its own policy of unification, albeit the policy may temporarily forestall military action by China.

Some liberal members of the KMT élite, such as the now-dissolved Wisdom Coalition and the newly formed Public Opinion faction (*Minyi hui*) in the Legislative Yuan, have urged the government to reconsider the 'one China' policy in order to break out of the diplomatic isolation imposed by Beijing. This has been stubbornly resisted by the conservative élite who consider it equivalent to pursuing Taiwan independence. Beijing also considers such proposals as 'separatist' and urges the KMT to enter into direct party-to-party negotiations with the CCP.

The Stands

The KMT

With only one exception, all members of the KMT élite interviewed by the author viewed unification as a long-term national goal, to be achieved regardless of the disparity in military power between the two sides of the Taiwan Strait. This point was stressed by President Lee Teng-hui in a written interview:

> Unification is the common desire of all Chinese people. Many people think that it would be very difficult for the Republic

of China to accomplish 'unification' in the current situation, because mainland China has a much larger surface area and the Chinese communists control more human and material power than we do. I personally do not agree with this, because this kind of argument overlooks the efforts and the influence of the people of the Republic of China and the overseas Chinese. It also overlooks the desire for and the pursuit of freedom, democracy, and prosperity of all Chinese. (written interview, 6 October 1988)

The government's unification policy has been modified significantly since the 1950s and 1960s. Chiang Kai-shek's only solution was a military takeover, and 'reconquering the mainland' (*fangong dalu*) was the government's standard slogan with regard to China at that time. As the former minister of economics and defense, Chen Li-an, revealed in an interview, after Chiang Ching-kuo emerged as the national leader in the early 1970s, strategic planning became defence-oriented, and in the slogans, 'reconquering' was gradually replaced by 'unification'.

From journal reports, there seems to be little consensus on the unification issue among KMT leaders. At one extreme, a very tiny minority of hardliners still hope for a military victory over the communist Chinese. Another very small minority at the other extreme are willing to unify with the communist regime under any circumstances (*The Journalist*, 81: 16–25). The irony is that many of those who want unification now were previously ardent anticommunists (*The Journalist*, 81: 29).

Nevertheless, most current members of the KMT élite are far from these two extremes. What they propose is basically a 'German model', two separate but equal states which recognize the possibility of future national unification. This is expressed by the Taipei government as 'one country, two central governments'. And in relation to its efforts in the early 1990s to rejoin international organizations such as the United Nations and the General Agreement on Tariffs and Trade (GATT), the government has proposed the 'one country, two seats' formula. During personal interviews, some members of the KMT élite stated that the objective of unification is not to replace the communist regime in Beijing, but to bring prosperity and democracy to mainland China.

One high-ranking government official claims to have been the earliest advocate of unification by non-military means:

In March 1979 . . . I made a report to the Central Standing Committee on the new relations with mainland China. In that report, I made two important points: First, we should talk about unification instead of reconquering the mainland. Unification would be a long-term goal. When I refer to unification, I mean something analogous to the long-term national goal of unification between East and West Germany. We do not need short-term policies to achieve that long-term goal. I was the first and the only person talking about 'unification' instead of reconquering the mainland at that time, and many disagreed with me. The second point I made was that we wanted the democratization and liberalization of mainland China, and we hoped for the good of the people there. In other words, we wanted to engage in peaceful competition with mainland China. This theme was hardly acceptable at the time. But I saw the views of our top leaders gradually change. (personal interview, 10 September 1988)

This official's view was echoed by others, one of whom stated that the object of unification with mainland China should be to bring democracy, freedom, and prosperity to the mainland. In a written interview, President Lee stated the official policy on the issue:

In recent years, the economic miracle in Taiwan created by the Republic of China has forced the communist leaders to raise the slogan 'learning economics from Taiwan' and to reveal their desire for accelerated and more profound economic reforms. When economic reform has developed to a certain stage, political reform will inevitably be encouraged. It will not be just superficial reform of political institutions, but reform that moves towards real liberal democracy.

Using rapid economic development to stimulate communist China to reform its economy is the first step the Republic of China should take in moving towards the goal of unification. The second step is to force communist China to carry out political reforms through our example of enforcing constitutional democracy, raising standards of education, and moving towards a pluralistic society. As long as our economic development is rapid and our political reform fast and realistic, there is no way that mainland China will not go in this direction, and there is no way that mainland China will not give up the 'four insistances'. As a result, the unification of China can be expected to come soon. (written interview, 6 October 1988)

This brief quotation from President Lee's response presents a rather blurry picture of unification based on political and economic reform in mainland China. This deliberate lack of specifics on how the goal can be achieved coincides with the statement in the previous interview that unification is a long-term goal which does not require short term policies. Thus, for these people, unification is only a spiritual goal.

No matter how remote are the chances of unification, all but one of the KMT interviewees recognized the need to retain it as a national goal. They are adamant that independence would provoke a disastrous attack from the mainland. Immediately after the Tiananmen massacre of 1989, the threat of military action by Beijing became very real. But yet again, events like Tiananmen might also encourage more people in Taiwan to favour independence in order to avoid communist rule.

The KMT's insistence on unification and its claim to represent the whole of China placed the government in a dilemma over democratization. In order to achieve democracy, Taiwan had to have a parliament and other institutions which reflected the will of the people of Taiwan. But this would sever Taiwan's legal and institutional bonds with China and thus make the government's claim to represent the Chinese people on the mainland untenable. Taiwan, with its own national symbol, its own constitution, its own set of political institutions, none of them linked to China, would have all the appearances of an independent country. Talk of unification in such circumstances would imply being taken over by the mainland. This dilemma kept the KMT government from pursuing wholehearted democratization.

As was pointed out above, the current ruling elité's conception of unification is quite vague. By pursuing a more active foreign policy, the government has injected some flexibility into the 'one China' principle. The 'one country, two central governments' model (or the 'German model', prior to reunification of the two Germanies) has been vigorously promoted by the liberal wing of the KMT since the party's Thirteenth Congress in the summer of 1988.

However, the policy of 'flexible diplomacy' was not adopted without considerable opposition from conservatives in the KMT. For example, during a meeting of the CSC on 12 October 1988, Shen Chang-huan, secretary-general of the presidential office, bitterly criticized the foreign minister, Lien Chan, and economics minister, Chen Li-an, for permitting a large group of government officials and business people to visit the Soviet Union. Shen had been the architect of the policy of refusal to coexist with mainland China in the international community (the so-called *hanzei bu liangli*) and was the most powerful foreign policy maker in the KMT. He was reportedly waving a copy of Chiang Kai-shek's book, *The Soviet Union in China*, at the time and nearly shouting at the younger ministers for not respecting the policy of anticommunism (*The Journalist*, 84: 12–22). This was part of a last-ditch effort by the conservatives to reverse the trend of reconciliation with communist countries including China. Public and media criticism of this incident and Shen's part in previous diplomatic failures was obviously too much for him; he submitted his resignation to Lee only a few days later and left the foreign policy-making arena for good (*The Journalist*, 85: 34–43).

One of the most important and the most striking examples of flexible diplomacy was the visit of Taiwan's finance minister, Shirley Kuo, to Beijing in early May 1989 to attend the annual meeting of the Asian Development Bank (*China Times*, 9 May 1989: 1). The visit destroyed the myth that the communist regime in China was Taiwan's arch enemy, and it re-cognized, albeit implicitly, the reality of communist rule in China. The trip was prepared in secret, and the sudden announcement took the country by surprise (*China Times*, 7 April 1989: 1). Even though the more conservative elements in the KMT were not happy about the visit, most people in Taiwan seemed to welcome the decision to break out of the diplomatic deadlock with China.

At a Central Committee Plenum in June 1989 Lee Teng-hui declared that Taiwan 'should have the courage to face the fact that it is not able to exercise effective rule over mainland China'. He added that Taiwan should not base its fate on an unrealistic claim, nor should Taiwan's diplomacy be restricted by its own ideology (*Central Daily News*, 3 June 1989: 1). These remarks departed sharply from the rigid line of the Chiang Kai-shek era and became the most important foundation for the policy of flexible diplomacy. Lee's remarks implied government recognition that the communist regime in Beijing was there to stay, no matter how unpalatable that fact was to the KMT government.

Although the KMT government tried to break out of the

diplomatic deadlock with China, it apparently found it difficult to extend this flexibility into the domestic arena. After the foreign minister, Lien Chan, had announced in the Legislative Yuan that the government was researching the feasibility of the 'one country, two central governments' model, the justice minister, Hsiao Tien-tzang said in reply to a questioning legislator that the government would consider rescinding the 'emergency decree' since it had recognized Beijing as another central government effectively controlling the mainland (*China Times*, 17 April 1989: 2). But a few days later, the government tried to explain that what Hsiao had said was no more than a slip of the tongue. This incident demonstrated that no matter how great a breakthrough Taiwan might achieve on the diplomatic front, domestic policies would be slow to be affected. It took a much more powerful driving force, such as the student demonstrations of March 1990, to push the government into action in the domestic arena.

After the 1990–91 crisis, the debate within the KMT over the national identity issue intensified and became the focal point of the power struggle. To Lee Teng-hui and those who supported him, democratization was seen as naturally leading to the 'Taiwanization' of the KMT, that is, powerful positions in the party and the government would be filled by an élite supported by the local population. As a result of this process, mainlanders—particularly older mainlanders who had never stood for election—would gradually lose their privileged position in politics and be replaced by Taiwanese. The fears of the mainlanders were compounded by the fact that the Mainstream faction's understanding on this point coincided with that of the DPP and that both realized that solidarity with the Chinese on the mainland had to come second to the welfare of the people of Taiwan. The mainlander-based Non-mainstream faction then accused President Lee of collaborating with the DPP to achieve independence, even though Lee had publicly stated on numerous occasions that the government would never entertain the idea of independence from China. The Non-mainstream faction charged that Lee's support for 'Taiwan independence in substance' was no different from the DPP's advocacy of Taiwan independence.

In order to quiet the protests of the Non-mainstream faction, Lee decided to form a supra-government institution, the National Unification Council (*Guojia tongyi weiyuanhui*), and issued the 'Guidelines for National Unification' (*Guojia Tongyi Gangling*) on 14 March 1991 as a formal statement of government policy towards mainland China. The guidelines proposed a three-phase process of unification and stated that 'after an appropriate period of forthright exchange, co-operation, and consultation conducted under the principles of reason, peace, parity, and reciprocity, the two sides of the Taiwan Strait should foster a consensus of democracy, free-dom and equal prosperity, and together build anew a unified China'. This was Lee's way of rebutting the attacks of the Non-mainstream faction.

The 'Guidelines for National Unification' were not enough to halt the attacks of the non-mainstreamers, however. Hau Pei-tsun, the former chief of the general staff who had been appointed premier in 1990, was considered by mainlanders to be the only figure powerful enough to hold back the tide of 'Taiwanization' and safeguard the position of mainlanders in the party. When Hau was being pressed to submit to reconfirmation by the Legislative Yuan after the general elec-tion in December 1992, leading Non-mainstream members organized rallies and demonstrations in front of the KMT head-quarters and Lee Teng-hui's residence. Some leading pro-testers even issued an open statement requesting Deng Xiaoping to watch developments in Taiwan closely and suggested in private that they would support Chinese troops in the event of an invasion. Another incident that illustrates the strength of feeling on both sides of the Taiwan-mainland debate occurred when Premier Hau was invited to deliver an executive report to the National Assembly at the end of 1992. When Hau stepped into the hall, many members of the Assembly, both DPP and KMT mainstreamers, shouted at Hau and called for him to step down. Responding to the heckling, Hau raised his arms and shouted 'Down with Taiwan independence' and 'Long live the Republic of China', and these slogans were echoed by the non-mainstreamers in the Assembly. Hau had become a symbol of conservatism and a stumbling block to Taiwan's political development in the eyes of the mainstreamers and the DPP, while the non-mainstreamers saw him as the only person capable of safeguarding the KMT's long-term national goal of unification. Faced with inevitable defeat in the legislature had he submitted to reconfirmation, Hau even-tually resigned and was replaced by the provincial governor,

Lien Chan, a trusted ally of Lee Teng-hui and an architect of 'flexible diplomacy' during his term as foreign minister.

The DPP

The dilemma faced by the DPP on the issue of unification is completely different from that of the KMT. From the author's interviews, it is clear that members of the New Tide faction want to see an independent Taiwan established immediately, while the leaders of the Formosa faction consider that a declaration of independence would alienate the DPP from the general public and ruin its chances of gaining power. Formosa believes that democratization should be the immediate goal for Taiwan, with the question of independence being determined in the future by the people.

In spite of this division, the two main factions agree that the KMT has used unification as an excuse to delay the process of democratization. They have no interest in the KMT's claim that the Nationalist government is the legitimate representative of the people on the Chinese mainland. The main concern of the majority of the opposition is that Taiwan should enter some kind of *rapprochement* with China in order to safeguard the island's prosperity and democracy. The opposition has argued that the KMT is not really interested in unification; it has merely used it to justify the party's continuing domination of society. Without unification as the ultimate goal, the KMT would have to give up its privileges and face the challenge of democracy. They claim that the government's 'three nos' policy of no contact, no negotiation, and no compromise with China demonstrates that the KMT does not really want unification, because the policy has actually created a *de facto* independent Taiwan.

In the eyes of some members of the opposition élite, unification with China could take only two forms: Taiwan would either be swallowed up by China or defeat China militarily. The latter, they protest, is impossible considering the discrepancy in size and military strength between the two sides, and the former would meet overwhelming resistance from most Taiwanese who would not wish to be ruled by a communist regime. Most of the interviewees agreed that the KMT's unrealistic claim to represent the Chinese people was the source of Taiwan's bizarre political system.

The response of You Ching, the Taipei county magistrate was typical:

> I don't think anyone would be so stupid as to agree with the argument that unification with China is a historical mission. Let me illustrate to you the reason why Taiwan should not unify with China. The average annual income of the people on mainland China is only about US$250, but that of the Taiwan people is about US$5,000 [approximate figures at the time of the interview], and China has one billion people and Taiwan only twenty million. What is Taiwan's annual income going to be when it is unified by China? Maybe less than US$500. If you think it is a good thing to reduce our annual income from $5,000 down to $500 you must have some serious mental problems. (personal interview, 26 August 1988)

Even though the DPP élite is unanimous in rejecting the notion of unification, some DPP leaders, particularly those who belong to the more moderate Formosa faction, are sensitive to the China factor and realize that hostile action by Beijing might cause irreparable damage to the prosperity and freedom of Taiwan. Lin Cheng-chieh, who withdrew from the DPP over the unification issue in mid-1991, pointed out that there are other solutions to Taiwan's conflict with China apart from the two extremes of unification or independence. To Lin, some kind of federation or confederation might both satisfy the ego of the Beijing regime and preserve the prosperity of Taiwan. Chang Chun-hong, a former secretary-general of the DPP and currently a legislator, also considers friendly relations with China as essential for Taiwan's survival. He argues that Taiwan cannot afford a war with a military giant; only friendly and co-operative relations, both economic and political, will serve the interests of both China and Taiwan. Again, You Ching's comments are worth noting:

> I would argue that we should talk about independence later, and only when we achieve democracy. . . . In the National Security Law, there is an article dealing with goods smuggled from China, and the rule is that the activity is treated just like smuggling from a foreign country. You can see that Taiwan and China are geographically separated by ocean, and we have laws recognizing China as another country. Therefore we do not have to worry about whether to pursue independence or not. . . . We do not have to be hostile to China. We should withdraw our troops from Kinmen and China should withdraw

from Xiamen year by year in order to reduce the hostility. We should also strengthen cultural and economic relations with each other to reduce the possibility of a military confrontation. (personal interview, 26 August 1988)

Some moderate DPP leaders believe that the blind pursuit of independence may do more harm than good to the opposition party and the people of Taiwan. Chang Chun-hong summed it up in the following passage:

An independent Taiwan is a very good idea, and it is the final solution to Taiwan's political problems. But the KMT, together with Communist China, is making the term a terrible concept to the people of Taiwan. The people here, including the KMT's own members, are all frightened by the KMT's interpretation of Taiwan independence. The meaning of an independent Taiwan has been twisted and it has become a symbol of terror. The KMT will not put you in prison now if you uphold the slogan of Taiwan independence, but it makes the middle class believe that if you put the concept into action, there will be turmoil in Taiwan and existing interests will be jeopardized. This has become a very complicated psychological problem. (personal interview, 6 September 1988)

Chang further pointed out that Taiwan independence is now 'poison' for the DPP, whereas for the KMT, it is no longer 'poison' but 'good food'. Judging from the DPP's poor performance in the 1991 National Assembly election, which is generally blamed on the party's adoption of an independence platform, Chang is quite correct.

Members of the New Tide élite unequivocally and unanimously argue that independence should be the immediate national goal for Taiwan. Chiou I-jen, a former deputy secretary-general of the DPP and New Tide's top ideologist, made the following statement during a personal interview:

It is necessary to have a thorough change of the government system and institutions in order to bring about democracy in Taiwan. Unfortunately, the problem of structural change in Taiwan is intermingled with the problem of national identity. The larger structural problems, such as parliamentary reform, the debate over a presidential or a parliamentary system, independence of the judiciary, etc., are all mixed up with the issue of national identity. This has not happened in any other Third World country. . . . We should therefore deal with the issue of structural change and the issue of national identity at the same time. If you only want to talk about structural change without touching on the issue of national identity, I think it is just not realistic. We have to think whether the new structure should be based on Taiwan only or whether the new structure should be based on China as a whole. (11 August 1988)

Chiou's remarks on Taiwan independence as a prerequisite for democracy were further articulated by Liu Feng-sung, another leader of the New Tide faction, who put it very strongly that there would be no democracy until Taiwan was declared independent from China. He argued, 'everything will be solved when Taiwan is declared independent', and 'it is impossible for the DPP to become a democratic party until then' (personal interview, 13 July 1988). The idea that the DPP must be a revolutionary party until Taiwan becomes independent is worrying to the general public. The moderates also view this position as a threat to the DPP's chances of gaining power and to democratization in general. Nevertheless, the radicals are very persistent. Fear of Taiwan independence, they say, is a result of KMT propaganda, and the DPP should launch its own propaganda campaign to explain to the people that independence is the only solution to Taiwan's mounting structural and diplomatic problems.

The moderates claim that the pursuit of independence is undemocratic because, in the present circumstances, the majority of people in Taiwan cannot accept the idea. The radicals, however, assert that there is no basic contradiction between the two ideas. Yao Chia-wen, who was the DPP's chairman at the time of this interview, made the following remarks concerning the two goals of the opposition:

I don't think you can really compare the two goals. . . . The greatest obstacle to Taiwan's democratization is the problem of national identity. If Taiwan is not independent, no matter how you reform, the parliament will still be a Chinese parliament, not a Taiwanese parliament. Therefore, without independence, full democratization is just a castle built on sand. . . . The reason why Taiwan has no democracy today is because the KMT government thinks that it is a Chinese government, not a Taiwanese government. In order to have democracy, we should recognize that we want a Taiwanese parliament. That is to say, Taiwan should be independent so that democracy can be achieved. (personal interview, 8 August 1988)

Taiwan independence might be consistent with democracy in some senses, as Yao suggested, but to pursue independence as a higher goal than democracy without considering the possible social consequences and China's reaction might very well jeopardize the future of democratization. The emotion attached to this issue may also lead to political polarization, and make democracy unattainable. Moreover, the radicals want independence regardless of the wishes of the electorate and believe that the way to achieve it is to use mass movements to overthrow the KMT government. It was Hong Chi-chang, another New Tide leader and a legislator, who said that he needed only 50,000 loyal followers to overthrow the KMT regime and build a new and independent Taiwan.[8]

The radicals' insistence on the use of Philippines-style 'people power' to overthrow the present government keeps the two opposition factions at odds with one another. Their distrust and distaste for each other has been displayed on many occasions, and it was only after Hsu Hsin-liang was elected DPP chairman that the two factions were able to reach a temporary compromise on the issue, agreeing that it would eventually be decided by a national plebiscite. However, national identity is still a potentially explosive issue among the opposition élite and between the KMT and the DPP.

The Prospects for a Majority Alliance

In the previous sections, the deep divisions among the political élite of both main parties on the key issues of parliamentary reform, constitutional revision, and national identity have been analysed. The issue of parliamentary reform was eventually resolved by retiring the mainland-elected parliamentarians and creating at-large seats, a compromise that satisfied both the conservatives and the reformers. On this issue, the reformers in the KMT, represented by Lee Teng-hui, used public support for their line to disarm the conservatives, forming an alliance with the student demonstrators and the entire DPP, including the most radical faction. Without the students and the DPP demonstrators of March 1990, Lee Teng-hui would have had little chance of getting his political reforms accepted by the KMT. The two other issues of constitutional revision and achieving a *modus vivendi* with China are still

Table 5.2 Party Composition of the Second National Assembly, 1993

Party	Contested	At-large	Overseas	Total	%
KMT	179	60	20	259	79.4
DPP	40[a]	17[b]	5	62	19.0
Independent	5[c]	0	0	5	1.6
Total	224	77	25	326	[100.0]

Notes: [a] 41 elected in 1991, but one has since died.
[b] 20 at-large seats in 1991 election, but three members were obliged to resign after winning Legislative Yuan seats in 1992.
[c] Includes three members of the Independents' Alliance formed prior to the 1991 election.
Source: *United Daily News*, 22 December, 1991: 1.

being debated, and further democratization, to a large degree, may once again depend on whether government and opposition leaders can put together a reform-oriented majority alliance or a compromise that can satisfy different forces.

A very important aspect of resolving the outstanding issues democratically—that is, by counting heads in the established representative bodies—is whether a clear reform-oriented majority can be formed. Constitutional revision is basically the responsibility of the National Assembly (which must also approve constitutional revision resolutions passed by the legislature), and any proposal to amend the current constitution needs a two-thirds majority in the Assembly.

In the National Assembly elected in December 1991, the KMT has nearly an 80 per cent majority (see Table 5.2), which should allow it to revise the constitution at will. However, the KMT cannot rely on its delegates to adopt a common stand, as was seen in the dispute over the key issue of the presidential election in 1992.

The KMT itself is evenly divided over how the president should be elected (see Figure 5.1). Those who support the Mainstream faction insist that according to democratic principles, the most important government position must be subject to

Figure 5.1 Groupings in the National Assembly

Party Affiliation

Faction		KMT	DPP
	Mainstream/ Moderates	Direct Election	Formosa
	Non- mainstream/ Radicals	Indirect Election Military	New Tide

direct popular election. Members of the Non-mainstream faction in the National Assembly, on the other hand, together with delegates with ties to the military, argue that direct election of the president would create an independent Taiwan and make unification a very dim prospect. The non-mainstreamers propose that the National Assembly be made into an electoral college on the American model. But now that the National Assembly has been turned into a powerful body, authorized to confirm various central government appointments and revise the constitution, it would be unrealistic to expect it to relegate itself to an electoral college role.

After Lee Teng-hui appointed his ally, Lien Chan, as premier in early 1993 and Lee himself was re-elected party chairman at the KMT's Fourteenth Congress in August that year, it became clear that there was a marked trend towards the president becoming the centre of government power, rather than the premier or the cabinet. Under these circumstances, a presidential system, with the president directly elected by the public, is necessary if the central government is to be democratic in essence. This is the line taken by the DPP as a whole, and they are likely to align with the KMT's Mainstream faction when the issue is decided sometime before the 20 May 1995 deadline. Such an alliance would give supporters of direct election the two-thirds majority needed to pass the amendment. This was the kind of alliance the DPP had hoped for during the 1992 constitutional revision process, but these hopes were dashed when the KMT dcided to shelve the issue for fear of splitting the party. But there is now a deadline for deciding the issue, and this will make the KMT's Mainstream faction seek allies in all directions in order to have its stand adopted in the National Assembly.

The Non-mainstream faction's position was further undermined when some of its younger members broke from the KMT to form the New Party in August 1993. At a press conference held to announce the formation of the New Party, Jaw Shao-kong, formerly the leader of the New KMT Alliance, said that a presidential system had become inevitable, and that it was necessary to have a directly elected president accompanicd by a proper system of checks and balances to safeguard democracy and prevent the president from becoming a dictator (*China News*, 11 August 1993: 2). If the New Party exerts its influence on the non-mainstreamers in the KMT or the non-mainstreamers respond to the call for direct presidential elections, such a system is even more likely to be adopted.

On the issue of national identity and relations with mainland China, a cross-party alliance between factions will be more difficult to achieve. One might expect that the KMT's Mainstream faction would ally with the DPP's Formosa faction, especially since Lee Teng-hui is an old associate of Hsu Hsin-liang and Chang Chun-hong from the 1970s when the two Formosa leaders ran the journal *University*, and he has had friendly relations with another leading Formosa figure, Huang Hsin-chieh, since 1990. However, neither the Non-mainstream faction of the KMT nor New Tide are likely to compromise their stands and allow 'independence in substance' to become the majority consensus. Moreover, there are strong emotions attached to this issue on both sides and any attempt to form an open alliance might provoke a sharp reaction from the two extremes and make the issue even harder to resolve.

Taking the response of the Legislative Yuan on the issue of reentering the United Nations as an example, four sets of stands are distinguishable among the four major groupings in the Yuan (see Figure 5.2). The KMT non-mainstreamers, including those legislators backed by the military, are not enthusiastic about the issue because they see the UN bid as a DPP ploy

Figure 5.2 Groupings in the Legislative Yuan

		Party Affiliation	
		KMT	DPP
Faction	Mainstream/ Moderates	Public Opinion Jade Mountain	Formosa Justice Alliance
	Non-mainstream/ Radicals	New KMT Alliance Militry	New Tide Welfare State Alliance WUFI

to create the impression that Taiwan is an independent country. The mainstreamers, including the twenty-member Public Opinion faction led by Kao Yu-jen and the fifteen-member Jade Mountain faction (Yushan hui) led by Tseng Chen-nung (which includes the surviving members of the disbanded Wisdom Coalition), began showing an interest in the issue only after the DPP had held several large rallies demanding the government seek a UN seat for Taiwan and President Lee made a positive response to their call. These mainstreamers insist, however, that Taiwan should apply for membership under the title 'Republic of China' and demonstrate to the international community that China is a divided nation like pre-unification Germany. The DPP as a whole is very interested in the issue of UN membership. The difference is that most Formosa legislators are not particular about the title under which the application is made, whereas New Tide, Welfare State Alliance, and the WUFI faction are pressing for the title to be decided by a plebiscite (they favour 'Republic of Taiwan'). These groups also consider UN membership to be an important step towards independence and one that would obviate the need for a formal declaration, which might anger the Chinese.

The KMT appears to have a two-to-one advantage over the DPP in the legislature (see Table 5.3), but because of the KMT legislators' low attendance rate and the party's factional

Table 5.3 Party Composition of the Legislative Yuan, 1993

Party	Contested	At-large	Overseas	Total	%
KMT	72[a]	19	4	95	59.0
DPP	39[a]	11	2	52	32.3
New Party	6	0	0	6	3.7
CSD[b]	1	0	0	1	.6
Independent	7[a]	0	0	7	4.3
Total	125	30	6	161	[99.9]

Notes: [a] These figures differ from the 1992 election results on account of one independent since joining the DPP and six members withdrawing from the KMT to form the New Party in September 1993.
[b] Chinese Social Democratic Party.
Source: *China Times*, 20 December 1992.

problems, the situation is in reality much more complicated. Despite this uncertainty, it is unlikely, for the time being at least, that moderates in the KMT and the DPP will form an alliance on the unification issue. The fact that the stage-by-stage unification process mapped out in Lee Teng-hui's 'Guidelines for National Unification' has been rejected by the DPP out of hand makes it quite obvious that the two sides cannot cooperate. Even though the Formosa faction is not in favour of immediate independence, it has never entertained the idea of unification. This is the gulf that divides the KMT's mainstreamers from the DPP moderates. And the Formosa faction is highly unlikely to risk the accusation that it is betraying Taiwan by changing its stance on this issue.

For the same reason, it would be very difficult for the KMT's mainstreamers to make a U-turn on policy towards mainland China. The 'Guidelines for National Unification', Lee's response to Non-mainstream attacks on his 'independence in substance' attitude, have already been criticized by non-mainstreamers who would like to see the immediate opening of direct communications with the mainland and negotiations with Beijing in the very near future. For Lee to push unification even farther into the future would only set the KMT's factional strife aflame again. Consequently, although many mainstreamers are in line with the DPP as far as priority for Taiwan is concerned, they can only ally with the DPP on side issues such as membership of the United Nations.

In the longer run, a compromise between the KMT's Mainstream faction and the DPP is not totally out of the question. Many of the arguments of the Wisdom Coalition concerning the need to focus on the political and economic development of Taiwan correspond to those of the DPP. Since the Wisdom Coalition's demise in the 1992 election, two new Mainstream groupings, Public Opinion and Jade Mountain, have taken up its standard, arguing that with regard to democratization and international relations, Taiwan should behave like a *de facto* state. In an open forum held in August 1993, Kao Yu-jen, head of the Public Opinion faction and a former member of the KMT's CSC, argued that Taiwan is in fact an independent country and said that he was opposed to the KMT's pursuit of unification with mainland China. Kao also suggested that the 'Guidelines for National Unification' should be converted into 'Guidelines for Cross-Strait Relations' (*Liberty Times*, 31 August 1993: 1). This kind of argument seems to make Mainstream legislators like Kao Yu-jen the natural allies of the DPP moderates on this issue.

However, the drawback of such a coalition is that mainstreamers in the executive branch are not ready to break away from the compromise they have reached with the conservatives, which is codified in the 'Guidelines for National Unification'. Moreover, from the results of the 1992 election, Mainstream legislators are likely to find it difficult to sell 'priority for Taiwan' to the electorate. Many election experts in Taiwan believe that the Wisdom Coalition's disastrous defeat in 1992 was due to the fact that it had blurred the line between the Mainstream KMT and the DPP on the unification issue and those voters persuaded by the 'priority for Taiwan' argument decided to vote for the DPP anyway, because its stand was more authentic and consistent. It seems that Mainstream legislators will have to risk election defeat if they try too hard to reach a compromise with the DPP moderates, making a majority coalition on this issue even more difficult to achieve.

Summary Conclusion

Taiwan's political élite exhibits a number of very different perceptions of democracy and how democracy can be achieved. The conflict of views and stands on the key political issues will likely make it very difficult to reach the kind of settlement necessary to build a new political system. These divisions also help to explain why the process of democratization begun in 1986 has been so prolonged, and why a democratic central government system—with either a head of state or chief executive directly elected by the people—is not likely to be in place before 1996.

The issue of national identity is probably the most controversial and the most fundamental problem that Taiwan has had to confront on the road to democracy. Other issues, such as parliamentary reform and the direct election of the president, are also complicated by the controversy over Taiwan's national status. At both extremes, this is an emotional issue which also involves the attitude of the Chinese government towards Taiwan, and for these reasons it is not likely to be resolved in the near future. In perhaps no other country in the world has the process of democratization depended so greatly on the attitude of an outside power. As Taiwan moves further along the road to democracy, the China factor will have to be addressed.

Chapter 6
∎

Conclusion: Prospects for Further Democratization

MANY changes took place in Taiwan during the seven years after President Chiang Ching-kuo announced his decision to undertake political reforms. Martial law was lifted, freedom of speech was safeguarded, opposition parties were legalized, the mainland-elected members of parliament were replaced, and the constitution was significantly revised.

However, viewed from an institutional perspective, Taiwan is still not a full democracy because neither the head of state nor the chief executive is elected by the people. The central government set-up—and the question whether power should be centred in the president or the premier—is still a subject of controversy. Concerning a free flow of information, the domestic electronic media are still largely controlled by the government and the ruling KMT.

Nevertheless, the public in Taiwan are generally optimistic about the prospects for democratization. Concerning the choice of a central government system, support for direct election of the president and a US-style presidential system seems to be mounting. A large group of legislators and National Assembly deputies have urged their respective chambers to adopt proposals allowing for the direct election of the president, and there is a growing movement in both the KMT and the DPP to ignore the provisions of the constitution regarding the term of the presidency and bring the presidential election forward. Whether this is done or not, there can be no doubt that resistance to a US-style system is fading. Although intense debate may be expected in the National Assembly prior to the May 1995 deadline, there is little doubt that the final vote will be in favour of direct election.

Once the presidential election issue is settled, the question will be what to do about the five-power central government setup, particularly the National Assembly. The Assembly, a

325-member parliamentary chamber whose chief functions are to elect the president and vice-president and to revise the constitution, will have very little purpose after the president is directly elected. Many people, especially those in the opposition camp, argue that the National Assembly is entirely unnecessary and should be abolished in order to save public money and avoid political controversy. But the irony is, of course, that a decision to abolish the Assembly, which would necessitate revision of the constitution, would have to be taken by the National Assembly itself. Assembly deputies, having spent so much money getting themselves elected, could hardly be expected to commit institutional suicide. What is more, during the last round of constitutional revisions, the National Assembly was given new powers to confirm the appointments of the presidents of the Examination Yuan and Judicial Yuan, the grand justices, and the members of the Control Yuan. Abolishing the National Assembly would therefore involve a complete overhaul of the constitutional provisions adopted in 1992. Intense public debate and institutional conflict between the Legislative Yuan and the National Assembly can be expected once again. But if the Assembly is preserved, public criticism against it is very likely to continue and this may lead to a constitutional crisis. To put it simply, the KMT-dominated second round of constitutional revision in 1992 tied a constitutional knot that will be very difficult to disentangle.

The social and economic forces behind democratization have been in process since the 1960s. Improved education, growing prosperity, and increasing contacts with the outside world have all played a part in the Taiwan people's quest for political rights, freedom, and participation. In the 1980s, the authoritarian regime came up against growing popular demand for more political freedom. The population was more defiant and street protests were frequent. Chiang Ching-kuo opted for reform chiefly because he sensed that further repression would cost the regime dear. He was farsighted enough to initiate reform at a time when the regime was still able to manage the pace of change, and he was the only individual in the regime with sufficient prestige to hold back the forces of conservatism.

As we have seen in Eastern Europe and elsewhere in the world, democratization is a juggernaut that cannot be stopped once it has been set in motion. In its efforts to democratize the polity at a manageable pace that would allow it to remain in power and maintain social order the KMT came up against both its own conservative wing and the opposition. The new reform-oriented KMT leadership then used public opinion to disarm the conservatives and push through parliamentary reform and constitutional revisions. In general, the pace of democratic transition has been quite slow compared to other countries, but the extent of the reforms is far-reaching and further change can be expected.

Despite the progress with democratization, the KMT's organizational strength and its strong patron-client relationship with local factions in the countryside are likely to make it a very difficult party to beat at the polls. Nevertheless, the KMT faces an entirely new situation in future elections. First of all, the DPP is growing stronger and better organized. Secondly, the KMT itself is riven with internal strife. The formation of the breakaway New Party may be just the beginning of the breakup of the ruling party. If the internal power struggle continues, more mainlander members of the KMT élite are likely to join or support the New Party, and the KMT will be weakened still further. In particular, if the local factions sense that the KMT is losing its grip on the distribution of political resources and it is no longer profitable for them to support the ruling party, they will put up their own candidates to compete against KMT nominees. Thus the factions' powerful networks of 'connections' will no longer be at the KMT's disposal and the ruling party will lose one of the pillars of its electoral success.

The question of national identity is still a highly controversial subject in Taiwan, and has influenced all aspects of democratization. Beijing's insistence that unification is the only solution to the conflict between the two sides of the Taiwan Strait and its oft-repeated threat to invade should Taiwan declare itself independent has added an element of fear to the debate. However, China's threats do not seem to bring unification any closer. On the contrary, there is a growing sense in Taiwan, and even within the KMT itself, that Taiwan should distance itself from the ruthless regime in Beijing and present itself in the international community as an independent political entity. Yet this heightened recognition that Taiwan does not and should not belong to China and that Taiwan should concentrate on its own political and economic development is not yet a national consensus. Indeed, opponents of any form of independence are threatening the unity of the KMT and the political stability of the country.

Where future democratization is concerned, the public is more worried about political order than about the establishment of a democratic political system. The KMT, as it painfully transforms itself from a quasi-Leninist to a democratic political party, is under increasing pressure in its effort to stay in power. The pressure is not only the result of intensified electoral competition, but also of factional conflict.

Along with the democratization of the polity in general come demands for the KMT to decentralize its structure. But the party's dilemma is that its sprawling organization, whose tentacles reach into every area of central and local government, the military, the education system, and the electronic media, as well as its own large and profitable business network, makes it very difficult for the KMT to decentralize without serious repercussions. People may want the KMT to retreat from society, but they also fear the chaos and disorder that might result. Meanwhile, there is a large segment of the party whose interests were guaranteed by the authoritarian ruling structure. These people fear that they will be deprived of their special privileges when the party and the polity are further democratized. They will do all they can to oppose the new leaders in the KMT who are spearheading these changes.

At the other end of the spectrum, the KMT leaders face challenges from their own elected representatives who are increasingly demanding a role in party policy-making. They argue that as popularly elected representatives of the people, they have more right to decide KMT policy than the top decision-makers, including the president, the premier, or the members of the Central Standing Committee, none of whom are popularly elected. In order to increase their leverage in the party, KMT representatives have formed their own factions and sub-groups. The leaders of these groups implicitly or explicitly declare themselves independent of KMT control, and are ready to pick and choose which KMT policies to support. The sub-groups in the Legislative Yuan, for example, leave the KMT center only about thirty-some legislators it can rely on, most of whom were selected by the party to occupy at-large seats. The serious defeat suffered by the KMT over the Office-holders' Personal Assets Disclosure Law, or 'sunshine law', in early 1993 provides ample illustration of how the KMT is losing control over its own ranks. This kind of internal dissention is threatening the KMT's status as the ruling party.

As if factional conflict were not enough, the KMT has been implicated in a number of well-substantiated corruption scandals in the 1990s, ranging from military procurement and highway construction to the purchase of computerized traffic signals and roadside parking meters in Taipei. The whiff of scandal has turned many voters against the ruling party. Furthermore, Lee Teng-hui's choice as premier, Lien Chan, is one of Taiwan's wealthiest men, and he long refused public calls to disclose his personal and family assets until forced to do so under the 'sunshine law'. The success of the New KMT Alliance in the 1992 legislative election was attributed by Alliance members to their strong criticism of corruption in the KMT and the leadership's close ties with big business. If the taint of corruption continues, the basis of the KMT's support will gradually be eroded.

In contrast to the troubled KMT, the DPP has grown much stronger since its establishment in September 1986. The 1992 Legislative Yuan election, in which the DPP won about one-third of both the popular vote and the total seats, took it one step closer to power. And in the 1993 election for county and city mayors, the DPP, despite its failure to gain any additional seats, received an unprecedented 41 per cent of the popular vote, only 6.5 percentage points behind the KMT. The opposition party's image has undergone a dramatic change, and it no longer engages in violence and disruption of normal business in the legislature. The absence of violence and disruption and the higher level of debate in the Legislative Yuan, plus greatly enhanced party unity, have changed the attitude of many people toward the DPP. Although many are still sceptical about the DPP's ability to lead the country, and some are worried that the coming to power of the DPP would provoke a Chinese attack, the opposition party is no longer regarded as a loosely-organized band of violent hotheads. In order to lay the ground for a bid for power, the DPP is aiming to capture the provincial governorship and win the Taipei and Kaohsiung mayoral races in 1994. The party then has its sights set on the presidential election scheduled for 1996.

Under the current political system, a party would need over

half the seats in the Legislative Yuan, or be able to form a
majority coalition, to influence the appointment of the premier and the cabinet. Outside the large urban centers, the electorate is still strongly influenced by factional, personal, and
kinship ties which work to the advantage of the KMT. So the
DPP is unlikely to win a majority in the legislature for the
time being, and indeed the party is not enthusiastic about
replacing the KMT and taking over the government through
this route. But should the National Assembly opt for an
American-style presidential system, this would work in the
DPP's favour as it would be much easier for the party to win
a one-on-one presidential race, especially now that the New
Party is likely to split the KMT vote.

However, the DPP's official line that Taiwan's status and
future should be decided solely by the people in Taiwan, and
the fact that many DPP leaders advocate a declaration of independence from China, is still considered seditious by some
people in Taiwan. This is the line taken by the Chinese government in Beijing, which in an August 1993 'white paper'
reiterated its warning that it will use force if Taiwan declares
itself independent. China's strong stand makes many people
in Taiwan hesitant to support the DPP, and this will be a serious obstacle for the party in future elections. The key DPP
leaders realize that public sentiment is against outright calls
for independence and they try to avoid discussion on the
issue. But it will take a few years before the DPP can rid itself
of its image as a 'Taiwan independence' party.

If there is one word which can sum up the prospects for
Taiwan in the future, it is 'uncertainty'. The KMT may or may
not lose power and the DPP may or may not replace it as the
ruling party. The KMT's internal power struggle may or may
not precipitate a final breakup. More seriously, the DPP coming to power may or may not provoke an attack from China.
People may look back nostalgically at the simple and pre-
dictable authoritarian order, but in doing so they will be 'over-
looking or regretting the transition's revival of precisely those
qualities which the previous regime has suppressed: creativity, hope, self-expression, solidarity, and freedom' (O'Donnell
and Schmitter, 1986: 4).

There is indeed almost unrestricted scope for creativity,
self-expression, and political freedom in Taiwan. All members of society can now hope to play a role in shaping the

future of the island and establishing a new political system. Politics is no longer dominated by one person or a small élite in the ruling party. The uncertainty is what people will have to get accustomed to, as there is no one at the top to give orders any longer. The process of democratization has already been set in motion, and it will not stop until a full democracy is realized.

Notes

Note to Chapter 1

[1.] Government domination of the electronic media is achieved through military, KMT, and provincial government shareholdings in the three television networks, plus a freeze on launching new television and radio stations. This system has been strongly criticized by the opposition, whose views have never been fairly represented in the media. Government control appears to be relaxing, however, and a bill legalizing cable television was passed by the Legislative Yuan in July 1993.

Notes to Chapter 2

[1.] This decree, the Temporary Provisions for the Period of General Mobilization to Suppress Rebellion (*Dongyuan kanluan shiqi linshi tiaokuan*), was enforced in 1948 at the height of the civil war between the Nationalists and the communists.

[2.] The confirmation procedure was changed in the 1992 constitution revision. The heads of the Examination Yuan and Judicial Yuan, along with members of the Control Yuan, are confirmed by the National Assembly.

[3.] The reason for the two ballots was that a candidate had to win the votes of over half the total membership of the National Assembly, which was 3,045. Failing that, a second ballot was held in which the winner required only a simple majority. The problem in 1954 was that the rump Assembly in Taiwan had only 1,573 members. To save himself from future embarrassment of this kind, Chiang had the rule changed by the Council of Grand Justices so that 'over half of the total votes' actually meant 'over half of the total able to be present'. See ROC Central Election Commission, *Summary of Elections in the Republic of China* (Taipei: CEC, 1984), Vol. 1: 14–20.

Notes to Chapter 3

[1.] Most of the leading figures involved in the journal have since become politically prominent. Hsu Hsin-liang, a leading editor, was later elected chairman of the DPP, and another editor, Chang Chun-hong, is a leading member of the opposition party. Both Hsu and Chang were KMT cadres at this time. Lee Teng-hui, a frequent visitor to the journal, became president of the Republic of China in 1988.

2. A lesser-known case of political persecution involved Pai Ya-tsan, a law school graduate who decided to run for a Legislative Yuan seat in central Taiwan in the 1975 election. He was arrested because his election flier contained a few sensitive questions addressed to Chiang Ching-kuo personally. These concerned the necessity of mar-tial law, the national goal of recovering mainland China, and par-liamentary reform. It appeared that Pai had asked more than Chiang could answer, and he was given a life sentence for sedition. See *The Journalist*, 59: 10.

3. December 10 is International Human Rights Day. The KMT and the opposition are still bitterly divided as to who started the viol-ence at the rally. Nevertheless, the riot gave the KMT an excuse to crack down on Dangwai activists.

4. Shih was pardoned at the time of Lee Teng-hui's inauguration in 1990. He was elected to the Legislative Yuan for his home town of Tainan in 1992 and became leader of the DPP caucus in the legis-lature. He became chairman of the DPP in 1994.

5. Issue no. 13 of *The Journalist* was devoted to the political prisoners of the *Formosa* group. It carried a detailed report on who was arrested, the length of each sentence, and where these people were at the moment. English-language reports of the incident can be found in *Far Eastern Economic Review* (FEER), 28 December 1979 and 25 April 1980.

6. The three murder cases were the killings of the twin daughters and mother of Lin Yi-hsiung, the death of Wang Wen-chen, and the murder of Henry Liu. Lin was a popular opposition member of the Provincial Assembly, Wang was a well-known dissident and a teacher at Carnegie Mellon University who died in mysterious circumstances while undergoing questioning by the Taiwan Garrison Command, and Liu was the author of a banned biography of Chiang Ching-kuo. Only the Liu murder was solved.

7. Ju later withdrew from the DPP and formed the Chinese Social Democratic Party. In 1993, Ju abandoned the party he had formed and joined the Chinese New Party.

8. The murder of Henry Liu was reported extensively by *The Asian Wall Street Journal* in the period between 16 October 1984 and 10 April 1985.

9. The financial crisis sparked by the loan scandal was reported extensively by *The Asian Wall Street Journal* on 15 February, 26 February, 12 March, and 15 May 1985.

10. According to government statistics, the average household income in Taiwan increased from NT$104,157 in 1964 to NT$309,901 in 1986 (at constant 1981 prices), while the ratio between the incomes of the top 20 per cent and the bottom 20 per cent fell from 5.33 to

4.60 in the same period. Government statistics also suggest that the majority of the population falls in the middle–upper income category. See *Directorate-General of BAS*, 1987a: 1, 6.

[11.] Prior to the 1992 revision of the constitution, in addition to the regular legislators elected by the general public, some legislators were elected by functional constituencies such as labour, farmers, and businessmen. In the 1986 election, Wu Yung-hsiung and Wang Tsung-sung, both DPP candidates, were elected to the Legislative Yuan as representatives of organized labour.

Notes to Chapter 4

[1.] Detailed result of the 1992 Legislative Yuan election were printed in *China Times*, 20 December 1992, p. 11.

[2.] The results of the 1986 National Assembly and Legislative Yuan elections are given in CEC, 1987.

[3.] According to the 1989 amendment of the Election Law, the six Legislative Yuan election districts in Taiwan Province were divided into smaller districts corresponding to county and provincial-level city boundaries. The two municipalities, Taipei and Kaohsiung, were given two districts each. For the 1991 National Assembly election, the Central Election Commission further divided each county and city into smaller independent districts with between four and seven seats in each.

[4.] Urban administrative units are, in order of size, *qu* (district), *li* (sub-division), and *lin* (neighbourhood). The rural units are *xian* (county), *zhen* and *xiang* (larger and smaller township), *cun* (village), and *lin* (neighbourhood).

[5.] For details of zoning and campaign activities in responsibility zones, see Chapter 3 of I-chou Liu, 'The Electoral Effect of Social Context Control on Voters: The Case of Taipei, Taiwan' (Ph.D. dissertation, University of Michigan, 1990).

[6.] Very often *zhuangjiao* are paid in cash for helping the candidate.

[7.] From a public speech by Hong Chi-chang, a DPP legislator and one of the top New Tide leaders, on 14 May 1989 in Columbus, Ohio.

[8.] The 1989 survey does not contain a question on the respondents' family income and therefore does not allow a more consistent comparison.

[9.] In contrast, an unpublished public opinion poll conducted by the Chinese Association of Political Science in 1993 indicated that age has an impact on attitude towards the government. Younger and better-educated respondents had a much stronger tendency than their elders to think that government officials are biased, corrupt, waste-ful of tax revenue, and therefore untrustworthy. If this represents a new trend in political attitude one may reasonably conclude that the KMT, as the ruling party, will gradually lose the support of younger and better-educated voters and the DPP will likely capitalize on the KMT's loss of support.

Notes to Chapter 5

1. This percentage may not seem high enough to constitute a pub-lic consensus on the issue, but considering that many people in Taiwan were afraid to express their political opinions after so many years of authoritarian rule and that the mass media in Taiwan was largely controlled by the government, the fact that over half of the population spoke out against government policy was considered a significant 'public consensus' at the time the survey was conducted.

2. *Laobiao*, a slang term of address used in Jiangxi Province, is, in this context, a shortened form of the term 'old representatives' (*lao daibiao*), used for the mainland-elected parliamentarians. It has a derogatory connotation in Taiwanese. The term *laozei*, old thieves, was first used by legislator Ju Gau-jeng.

3. The National Affairs Conference took place 27 June–4 July 1990, three months after the student protests. There are numerous news reports in Chinese on the conference, including analyses in the weekly news magazine *The Journalist*. In depth academic analyses may be found in a special issue of *Montai to Kenky* (Issues and Studies) (Tokyo) 20, No. 1 (October 1990).

4. This section on the ROC constitution draws substantially on an article published by the author in *Issues & Studies* (Taipei) 28, No. 9 (September 1982): 85–106.

5. The most serious conflict came on 16 April when a fist fight broke out between members of the two parties in which three people were injured. See *China Times*, 17 April 1992: 1, 2.

6. One technical twist in the questionnaire is that unification by the Taiwan government is only one form of unification. Other forms include unification by the PRC, federation, confederation, or any peaceful settlement reached between Taiwan and the PRC other than independence. As a result of this technicality, the percentage of those believing in unification might have been higher than the survey would imply.

7. The above data were supplied by the Public Opinion Research Foundation, Taipei.

8. A public speech delivered by Hong Chi-chang on 14 May 1989 in Columbus, Ohio.

Appendix I

Members of Political Élites Interviewed by the Author

Ruling Élite

Dr Lee Teng-hui
President of the Republic of China and chairman of the KMT.
President Lee received his Ph.D. degree from Cornell University.
He was a trusted aide of President Chiang Ching-kuo, and his
previous appointments include Taipei City mayor, provincial
governor, and vice-president. He succeeded to the presidency
on 13 January 1988 when Chiang died.

Dr Shih Chi-yang
Secretary-general of the National Security Council, and member of the KMT Central Standing Committee and the National
Assembly. Dr Shih earned his law degree in West Germany,
and formerly held the positions of minister of justice and vice-premier.

Dr Chen Li-an
President of the Control Yuan. He is the son of a former vice-president, Chen Cheng, who directed the land reform program
of the early 1950s. His previous positions include deputy
secretary-general of the KMT, head of the National Science
Council, minister of economics, and minister of defense.

Dr Ma Ying-jeou
Minister of Justice. He formerly served as deputy secretary-general of the KMT, chairman of the Research, Development,
and Evaluation Commission of the Executive Yuan, and vice-chairman of the Mainland Affairs Council. Dr Ma received his
law degree from Harvard University.

Dr Fredrick F. Chien
Minister of Foreign Affairs. He was formerly the ROC representative in Washington, DC, and chairman of the Council for

Economic Planning and Development. He is widely respected as one of the leading second-generation mainlanders in Taiwan. He received his Ph.D. in political science from Yale University, and is believed to be one of the most important architects of current Taiwan–US relations.

Dr Jeanne Chong-koei Li
Member of the KMT Central Standing Committee and deputy secretary-general of the party. She was formerly president of the China Youth Corps and head of the KMT Department of Women's Affairs. Dr Li received her law degree from the University of Paris.

Dr Chang King-yuh
Minister without portfolio and a member of the KMT Central Committee. He was formerly president of National Chengchi University and director of the Government Information Office and the government spokesman. Dr Chang received his Ph.D. in international relations from Columbia University.

Mr Lin Yu-hsiang
Former member of the Legislative Yuan and Taipei City Council. He was the top leader of the Wisdom Coalition and a member of the Mainstream faction.

Mr Hsu Hsin-chih
Member of the Control Yuan. He is a former deputy minister of interior and consultant to the Executive Yuan. He worked up the political ladder from the town government level, and therefore is more familiar with local government than most central government officials. Mr Hsu is considered to be a member of the Non-mainstream faction for his close association with Lin Yang-kang, president of the Judicial Yuan.

Mr Chu Po-chun
Former secretary-general of the Central Election Commission and a technical adviser to the Ministry of the Interior. Mr Chu is a political science professor at National Chengchi University.

Mr Chang Chao-chuan
Member of the Taiwan Provincial Assembly and a leader of the White faction in Changhua County.

Mr Tseng Teh-yun
Chief of Section One, Taichung county branch of the Nationalist

Party. Mr Chang's chief responsibility is the coordination between the party and the factions in the county.

Mr Hwang Chun-hsiung
Chief of Section One, Changhua county branch of the Nationalist Party. Mr Hwang's chief responsibility is the coordination between the party and the factions in the county.

Opposition Élite

Mr Yao Chia-wen
Elected to the Legislative Yuan in 1992. He was the chairman of the DPP when interviewed by the author.

Mr Chiou I-jen
Deputy secretary-general of the DPP, and a leader of the New Tide faction.

Dr You Ching
Magistrate of Taipei County since 1989. He is a former member of the Control Yuan and the Legislative Yuan.

Mr Chang Chun-hung
Member of the Legislative Yuan since 1992. He is a respected theoretician of the DPP. Mr Chang is one of the top leaders of the Formosa faction.

Mr Kang Ning-hsiang
Member of the Control Yuan since 1993. He is a former member of the Legislative Yuan and the Taipei City Council. He was the top leader of the opposition movement for a few years after the 1979 Kaohsiung incident.

Mr Fei Hsi-ping
Formerly a senior member of the Legislative Yuan and one of the founding fathers of the DPP. One of the few DPP leaders of mainland origin, Mr Fei withdrew from the party in December 1988 to protest against the DPP's position on the retirement of senior parliamentarians.

Ms Hsu Jong-shu
Member of the National Assembly. Ms Hsu is a former member of the Legislative Yuan and a leader of the Formosa faction.

Mr Lin Cheng-chieh
Member of the Legislative Yuan since 1989. He was formerly a member of Taipei City Council and the prime leader of the Progress faction of the DPP. He withdrew from the DPP in 1991 over issue of Taiwan independence.

Ms Weng Chin-chu
Member of the Legislative Yuan since 1992 and a leader of the New Tide faction. Ms Weng was a member of the National Assembly prior to 1992.

Mr Liu Feng-sung
An opposition writer and a leader of the New Tide faction. He is a member of the DPP Central Executive Committee.

Mr Hsu Mu-yuan
Member of the Taipei City Council and the Chief Executive of FAPA (Formosans' Association for Public Affairs) in Taiwan.

Mr Huang Shih-cheng
Minister without portfolio. Formerly the magistrate of Changhua County. Mr Huang does not have any party affiliation, but has warm relations with both the KMT and the DPP.

Mr Lee Sen-fong
Top leader of the Chinese New Party. He was elected into the Legislative Yuan in 1986, and became a leader of the New KMT Alliance and the Non-mainstream faction in 1989.

Appendix II

Constitution of the Republic of China

(Adopted by the National Assembly on 25 December 1946, promulgated by the National Government on 1 January 1947, and effective from 25 December 1947)

The National Assembly of the Republic of China, by virtue of the mandate received from the whole body of citizens, in accordance with the teachings bequeathed by Dr Sun Yat-sen in founding the Republic of China, and in order to consolidate the authority of the State, safeguard the rights of the people, ensure social tranquility, and promote the welfare of the people, do hereby establish this Constitution, to be promulgated throughout the country for faithful and perpetual observance by all.

Chapter I. General Provisions

Article 1. The Republic of China, founded on the Three Principles of the People, shall be a democratic republic of the people, to be governed by the people and for the people.

Article 2. The sovereignty of the Republic of China shall reside in the whole body of citizens.

Article 3. Persons possessing the nationality of the Republic of China shall be citizens of the Republic of China.

Article 4. The territory of the Republic of China according to its existing national boundaries shall not be altered except by resolution of the National Assembly.

Article 5. There shall be equality among the various racial groups in the Republic of China.

Article 6. The national flag of the Republic of China shall be of red ground with a blue sky and a white sun in the upper left corner.

Chapter II. Rights and Duties of the People

Article. 7 All citizens of the Republic of China, irrespective of sex, religion, race, class, or party affiliation, shall be equal before the law.

Article 8. Personal freedom shall be guaranteed to the people. Except in case of *flagrante delicto* as provided by law, no person shall be arrested or detained otherwise than by a judicial or a police organ in accordance with the procedure prescribed by law. No person shall be tried or punished otherwise than by a law court in accordance with the procedure prescribed by law. Any arrest, detention, trial, or punishment which is not in accordance with the procedure prescribed by law may be resisted.

When a person is arrested or detained on suspicion of having committed a crime, the organ making the arrest or detention shall in writing inform the said person, and his designated relative or friend, of the grounds for his arrest or competent court for trial. The said person, or any other person, may petition the competent court that a writ be served within 24 hours on the organ making the arrest for the surrender of the said person for trial.

The court shall not reject the petition mentioned in the preceding paragraph, nor shall it order the organ concerned to make an investigation and report first. The organ concerned shall not refuse to execute, or delay in executing, the writ of the court for the surrender of the said person for trial.

When a person is unlawfully arrested or detained by any organ, he or any other person may petition the court for an investigation. The court shall not reject such a petition, and shall, within 24 hours, investigate the action of the organ concerned and deal with the matter in accordance with law.

Article 9. Except for those in active military service, no person shall be subject to trial by a military tribunal.

Article 10. The people shall have freedom of residence and of change of residence.

Article 11. The people shall have freedom of speech, teaching, writing and publication.

Article 12. The people shall have freedom of privacy of correspondence.

Article 13. The people shall have freedom of religious belief.

Article 14. The people shall have freedom of assembly and association.

Article 15. The right of existence, the right of work and the right of property shall be guaranteed to the people.

Article 16. The people shall have the right of presenting petitions, lodging complaints, or instituting legal proceedings.

Article 17. The people shall have the rights of election, recall, initiative and referendum.

Article 18. The people shall have the rights of taking public examinations and of holding public offices.

Article 19. The people shall have the duty of paying taxes in accordance with law.

Article 20. The people shall have the duty of performing military service in accordance with law.

Article 21. The people shall have the right and the duty of receiving citizens' education.

Article 22. All other freedoms and rights of the people that are not detrimental to social order or public welfare shall be guaranteed under the Constitution.

Article 23. All the freedoms and rights enumerated in the preceding Article shall not be restricted by law except by such as may be necessary to prevent infringement upon the freedoms of other persons, to avert an imminent crisis, to maintain social order or to advance public welfare.

Article 24. Any public functionary who, in violation of law, infringes upon the freedom or right of any person shall, in addition to being subject to disciplinary measures in accordance with law, be held responsible under criminal and civil laws. The injured person may, in accordance with law, claim compensation from the State for damage sustained.

Chapter III. The National Assembly

Article 25. The National Assembly shall, in accordance with the provisions of this Constitution, exercise political powers on behalf of the whole body of citizens.

Article 26. The National Assembly shall be composed of the following delegates:

1. One delegate shall be elected from each *hsien*, municipality, or area of equivalent status. In case its population exceeds 500,000, one additional delegate shall be elected for each additional 500,000. Areas equivalent to *hsien* or municipalities shall be prescribed by law.

2. Delegates to represent Mongolia shall be elected on the basis of four for each league and one for each special banner.

3. The number of delegates to be elected from Tibet shall be prescribed by law.

4. The number of delegates to be elected by various racial groups in frontier regions shall be prescribed by law.

5. The number of delegates to be elected by Chinese citizens residing abroad shall be prescribed by law.

6. The number of delegates to be elected by occupational groups shall be prescribed by law.

7. The number of delegates to be elected by women's organizations shall be prescribed by law.

Article 27. The function of the National Assembly shall be as follows:

1. To elect the President and the Vice-President;

2. The recall the President and the Vice-President;

3. The amend the Constitution; and

4. To vote on proposed constitutional amendments submitted by the Legislative Yuan by way of referendum.

With respect to the rights of initiative and referendum, except as is provided in Items 3 and 4 of the preceding paragraph, the National Assembly shall make regulations pertaining thereto and put them into effect, after the above-mentioned two political rights shall have been exercised in one half of the *hsien* and municipalities of the whole country.

Article 28. Delegates to the National Assembly shall be elected every six years.

The term of office of the delegates to each National Assembly shall terminate on the day on which the next National Assembly convenes.

No incumbent government official shall, in the electoral area where he holds office, be elected delegate to the National Assembly.

Article 29. The National Assembly shall be convoked by the President to meet 90 days prior to the date of expiration of each presidential term.

Article 30. An extraordinary session of the National Assembly shall be convoked in any of the following circumstances:

1. When, in accordance with the provisions of Article 49 of this Constitution, a new President and a new Vice-President are to be elected;

2. When, by resolution of the Control Yuan, an impeachment of the President or the Vice-President is instituted;

3. When, by resolution of the Legislative Yuan, an amendment to the constitution is proposed; and

4. When a meeting is requested by not less than two-fifths of the delegates to the National Assembly.

When an extraordinary session is to be convoked in accordance with Item 1 or Item 2 of the preceding paragraph, the President of the Legislative Yuan shall issue the notice of convocation; when it is to be convoked in accordance with Item 3 or Item 4, it shall be convoked by the President of the Republic.

Article 31. The National Assembly shall meet at the seat of the Central Government.

Article 32. No delegate to the National Assembly shall be held responsible outside the Assembly for opinions expressed or votes cast at meetings of the Assembly.

Article 33. While the Assembly is in session, no delegate to the National Assembly shall, except in case of *flagrante delicto*, be arrested or detained without the permission of the National Assembly.

Article 34. The organization of the National Assembly, the election and recall of delegates to the National Assembly, and the procedure whereby the National Assembly is to carry out its functions, shall be prescribed by law.

Chapter IV. The President

Article 35. The President shall be the head of the State and shall represent the Republic of China in foreign relations.

Article 36. The President shall have supreme command of the land, sea and air forces of the whole country.

Article 37. The President shall, in accordance with law, promulgate laws and issue mandates with the counter-signature of the President of the Executive Yuan or with the counter-signatures of both the President of the Executive Yuan and the Ministers or Chairmen of Commissions concerned.

Article 38. The President shall, in accordance with the provisions of this Constitution, exercise the powers of concluding treaties, declaring war and making peace.

Article 39. The President may, in accordance with law, declare martial law with the approval of, or subject to confirmation by, the Legislative Yuan. When the Legislative Yuan deems it necessary, it may by resolution request the President to terminate martial law.

Article 40. The President shall, in accordance with law, exercise the power of granting amnesties, pardons, remission of sentences and restitution of civil rights.

Article 41. The President shall, in accordance with law, appoint and remove civil and military officials.

Article 42. The President may, in accordance with law, confer honours and decorations.

Article 43. In case of a natural calamity, an epidemic, or a national financial or economic crisis that calls for emergency measures, the President, during the recess of the Legislative Yuan, may, by resolution of the Executive Yuan Council, and in accordance with the Law on Emergency Orders, issue emergency orders, proclaiming such measures as may be necessary to cope with the situation. Such orders shall, within one month after issuance, be presented to the Legislative Yuan for confirmation; in case the Legislative Yuan withholds confirmation, the said orders shall forthwith cease to be valid.

Article 44. In case of disputes between two or more Yuan other than those concerning which there are relevant provisions in this Constitution, the President may call a meeting of the Presidents of the Yuan concerned for consultation with a view to reaching a solution.

Article 45. Any citizen of the Republic of China who has attained the age of 40 years may be elected President or Vice-President.

Article 46. The election of the President and the Vice-President shall be prescribed by law.

Article 47. The President and the Vice-President shall serve a term of six years. They may be re-elected for a second term.

Article 48. The President shall, at the time of assuming office, take the following oath:

'I do solemnly and sincerely swear before the people of the whole country that I will observe the Constitution, faithfully perform my duties, promote the welfare of the people, safeguard the security of the State, and will in no way betray the people's trust. Should I break my oath, I shall be willing to submit myself to severe punishment by the State. This is my solemn oath.'

Article 49. In case the office of the President should become vacant, the Vice-President shall succeed until the expiration of the original presidential term. In case the office of both the President and the Vice-President should become vacant, the President of the Executive Yuan shall act for the President; and, in accordance with the provisions of Article 30 of this Constitution, an extraordinary session of the National Assembly shall be convoked for the election of a new President and a new Vice-President, who shall hold office until the completion of the term left unfinished by the preceding President. In case the President should be unable to attend to office due to any cause, the Vice-President shall act for the President. In case both the President and Vice-President should be unable to attend to office, the President of the Executive Yuan shall act for the President.

Article 50. The President shall be relieved of his functions of the day on which his term of office expires. If by that time the succeeding President has not yet been elected, or if the President-elect and the Vice-President-elect have not yet assumed office, the President of the Executive Yuan shall act for the President.

Article 51. The period during which the President of the Executive Yuan may act for the President shall not exceed three months.

Article 52. The President shall not, without having been recalled, or having been relieved of his functions, be liable to criminal prosecution unless he is charged with having committed an act of rebellion or treason.

Chapter V. Administration

Article 53. The Executive Yuan shall be the highest administrative organ of the State.

Article 54. The Executive Yuan shall have a President, a Vice-President, a certain number of Ministers and Chairmen of commissions, and a certain number of Ministers without portfolio.

Article 55. The President of the Executive Yuan shall be nominated and, with the consent of the Legislative Yuan, appointed by the President of the Republic.

If, during the recess of the Legislative Yuan, the President of the Executive Yuan should resign or if his office should become vacant, his functions shall be exercised by the Vice-President of the Yuan, acting on his behalf, but the President of the Republic shall, within 40 days, request a meeting of the Legislative Yuan to confirm his nominee for the vacancy. Pending such confirmation, the Vice-President of the Executive Yuan shall temporarily exercise the functions of the President of the said Yuan.

Article 56. The Vice-President of the Executive Yuan, Ministers and Chairmen of Commissions, and Ministers Without Portfolio shall be appointed by the President of the Republic upon the recommendation of the President of the Executive Yuan.

Article 57. The Executive Yuan shall be responsible to the Legislative Yuan in accordance with the following provisions:

1. The Executive Yuan has the duty to present to the Legislative Yuan a statement of its administrative policies and a report on its administration. While the Legislative Yuan is in session, members of the Legislative Yuan shall have the right to question the President and the Ministers and Chairmen of Commissions of the Executive Yuan.

2. If the Legislative Yuan does not concur in any important policy of the Executive Yuan, it may, by resolution, request the Executive Yuan to alter such a policy. With respect to such resolution, the Executive Yuan may, with the approval of the President of the Republic, request the Legislative Yuan

for reconsideration. If, after reconsideration, two-thirds of the members of the Legislative Yuan present at the meeting uphold the original resolution, the President of the Executive Yuan shall either abide by the same or resign from office.

3. If the Executive Yuan deems a resolution on a statutory, budgetary, or treaty bill passed by the Legislative Yuan difficult of execution, it may, with the approval of the President of the Republic and within ten days after its transmission to the Executive Yuan, request the Legislative Yuan to reconsider the said resolution. If after reconsideration, two-thirds of the Legislative Yuan present at the meeting uphold the original resolution, the President of the Executive Yuan shall either abide by the same or resign from office.

Article 58. The Executive Yuan shall have an Executive Yuan Council, to be composed of its President, Vice-President, various Ministers and Chairmen of Commissions, and Ministers without Portfolio, with its President as Chairman.

Statutory or budgetary bills or bills concerning martial law, amnesty, declaration of war, conclusion of peace or treaties, and other important affairs, all of which are to be submitted to the Legislative Yuan, as well as matters that are of common concern to the various Ministries and Commissions, shall be presented by the President and various Ministers and Chairmen of Commissions of the Executive Yuan to the Executive Yuan Council for decision.

Article 59. The Executive Yuan shall, three months before the beginning of each fiscal year, present to the Legislative Yuan the budgetary bill for the following fiscal year.

Article 60. The Executive Yuan shall, within four months after the end of each fiscal year, present final accounts of revenues and expenditures to the Control Yuan.

Article 61. The organization of the Executive Yuan shall be prescribed by law.

Chapter VI. Legislation

Article 62. The Legislative Yuan shall be the highest legislative organ of the State, to be constituted of members elected by the people. It shall exercise legislative power on behalf of the people.

Article 63. The Legislative Yuan shall have the power to decide by resolution upon statutory or budgetary bills or bills concerning martial law, amnesty, declaration or war, conclusion of peace or treaties, and other important affairs of the State.

Article 64. Members of the Legislative Yuan shall be elected in accordance with the following provisions:

1. Those to be elected from the provinces and by the municipalities under the direct jurisdiction of the Executive Yuan shall be five for each province or municipality with a population of not more than 3,000,000, one additional member shall be elected for each additional 1,000,000 in a province or municipality whose population is over 3,000,000;

2. Those to be elected from Mongolian Leagues and Banners;

3. Those to be elected from Tibet;

4. Those to be elected by various racial groups in frontier regions;

5. Those to be elected by Chinese citizens residing abroad; and

6. Those to be elected by occupational groups.

The election of members of the Legislative Yuan and the number of those to be elected in accordance with Items 2 to 6 of the preceding paragraph shall be prescribed by law. The number of women to be elected under the various items enumerated in the first paragraph shall be prescribed by law.

Article 65. Members of the Legislative Yuan shall serve a term of three years, and shall be re-eligible. The election of members of the Legislative Yuan shall be completed within three months prior to the expiration of each term.

Article 66. The Legislative Yuan shall have a President and a Vice-President, who shall be elected by and from among its members.

Article 67. The Legislative Yuan may set up various committees. Such committees may invite government officials and private persons concerned to be present at their meetings to answer questions.

Article 68. The Legislative Yuan shall hold two sessions each year, and shall convene of its own accord. The first session shall last from February to the end of May, and the second session from September to the end of December. Whenever necessary, a session may be prolonged.

Article 69. In any of the following circumstances, the Legislative Yuan may hold an extraordinary session:

1. At the request of the President of the Republic; and
2. Upon the request of not less than one-fourth of its members.

Article 70. The Legislative Yuan shall not make proposals for an increase in the expenditures in the budgetary bill presented by the Executive Yuan.

Article 71. At the meetings of the Legislative Yuan, the Presidents of the various Yuan concerned and the various Ministers and Chairmen of Commissions concerned may be present to give their views.

Article 72. Statutory bills passed by the Legislative Yuan shall be transmitted to the President of the Republic and to the Executive Yuan. The President shall, within ten days after receipt thereof, promulgate them; or he may deal with them in accordance with the provisions of Article 57 of this Constitution.

Article 73. No member of the Legislative Yuan shall be held responsible outside the Yuan for opinions expressed or votes cast in the Yuan.

Article 74. No member of the Legislative Yuan shall, except in case of *flagrante delicto*, be arrested or detained without the permission of the Legislative Yuan.

Article 75. No member of the Legislative Yuan shall concurrently hold a government post.

Article 76. The organization of the Legislative Yuan shall be prescribed by law.

Chapter VII. Judiciary

Article 77. The Judicial Yuan shall be the highest judicial organ of the State and shall have charge of civil, criminal, and administrative cases, and over cases concerning disciplinary measures against public functionaries.

Article 78. The Judicial Yuan shall interpret the Constitution and shall have the power to unify the interpretation of laws and orders.

Article 79. The Judicial Yuan shall have a President and a Vice-President, who shall be nominated and, with the consent of the Control Yuan, appointed by the President of the Republic.

The Judicial Yuan shall have a certain number of Grand Justices to take charge of matters specified in Article 78 of this Constitution, who shall be nominated and, with the consent of the Control Yuan, appointed by the President of the Republic.

Article 80. Judges shall be above partisanship and shall, in accordance with law, hold trials independently, free from any interference.

Article 81. Judges shall hold office for life. No judge shall be removed from office unless he has been found guilty of a criminal offense or subjected to disciplinary measure, or declared to be under interdiction. No judge shall, except in accordance with law, be suspended or transferred or have his salary reduced.

Article 82. The organization of the Judicial Yuan and of the law courts of various grades shall be prescribed by law.

Chapter VIII. Examination

Article 83. The Examination Yuan shall be the highest examination organ of the State and shall have charge of matters relating to examination, employment, registration, service rating, scale of salaries, promotion and transfer, security of tenure, commendation, pecuniary aid in case of death, retirement and old age pension.

Article 84. The Examination Yuan shall have a President and a Vice-President and a certain number of members, all of whom shall be nominated and, with the consent of the Control Yuan, appointed by the President of the Republic.

Article 85. In the selection of public functionaries, a system of open competitive examination shall be put into operation, and examinations shall be held in different areas, with prescribed numbers of persons to be selected according to various provinces and areas. No person shall be appointed to a public office unless he is qualified through examination.

Article 86. The following qualifications shall be determined and registered through examination by the Examination Yuan in accordance with law:

1. Qualification for appointment as public functionaries; and
2. Qualification for practice in specialized professions or as technicians.

Article 87. The Examination Yuan may, with respect to matters under its charge, present statutory bills to the Legislative Yuan.

Article 88. Members of the Examination Yuan shall be above partisanship and shall independently exercise their functions in accordance with law.

Article 89. The organization of the Examination Yuan shall be prescribed by law.

Chapter IX. Control

Article 90. The Control Yuan shall be the highest control organ of the State and shall exercise the powers of consent, impeachment, censure and auditing.

Article 91. The Control Yuan shall be composed of members who shall be elected by Provincial and Municipal Councils, the local Councils of Mongolia and Tibet, and Chinese citizens residing abroad. Their numbers shall be determined in accordance with the following provisions:

1. Five members from each province;
2. Two members from each municipality under the direct jurisdiction of the Executive Yuan;
3. Eight members from Mongolian Leagues and Banners;
4. Eight members from Tibet; and
5. Eight members from Chinese citizens residing abroad.

Article 92. The Control Yuan shall have a President and a Vice-President, who shall be elected by and from among its members.

Article 93. Members of the Control Yuan shall serve a term of six years and shall be re-eligible.

Article 94. When the Control Yuan exercises the power of consent in accordance with this Constitution, it shall do so by resolution of a majority of the members present at the meeting.

Article 95. The Control Yuan may, in the exercise of its powers of control, request the Executive Yuan and its Ministries and Commissions to submit to it for perusal the original orders issued by them and all other relevant documents.

Article 96. The Control Yuan may, taking into account the work of the Executive Yuan and its various Ministries and Commissions, set up a certain number of committees to investigate their activities with a view to ascertaining whether or not they are guilty of violation of law or neglect of duty.

Article 97. The Control Yuan may, on the basis of the investigations and resolutions of its committees, propose corrective measures and forward them to the Executive Yuan and the Ministries and Commissions concerned, directing their attention to effecting improvements.

When the Control Yuan deems a public functionary in the Central Government or in a local government guilty of neglect of duty or violation of law, it may propose corrective measures or institute an impeachment. If it involves a criminal offense, the case shall be turned over to a law court.

Article 98. Impeachment by the Control Yuan of a public functionary in the Central Government or in a local government shall be instituted upon the proposal of one or more than one member of the Control Yuan and the decision, after due consideration, by a committee composed of not less than nine members.

Article 99. In case of impeachment by the Control Yuan of the personnel of the Judicial Yuan or of the Examination Yuan for neglect of duty or violation of law, the provisions of Articles 95, 97 and 98 of this Constitution shall be applicable.

Article 100. Impeachment by the Control Yuan of the President or the Vice-President of the Republic shall be instituted upon the proposal of not less than one-fourth of the whole body of members of the Control Yuan, and the resolution, after due consideration, by the majority of the whole body of members of the Control Yuan, and the same shall be presented to the National Assembly.

Article 101. No member of the Control Yuan shall be held responsible outside the Yuan for opinions expressed or votes cast in the Yuan.

Article 102. No member of the Control Yuan shall, except in case of *flagrante delicto*, be arrested or detained without the permission of the Control Yuan.

Article 103. No member of the Control Yuan shall concurrently hold a public office or engage in any profession.

Article 104. In the Control Yuan, there shall be an Auditor General who shall be nominated and, with the consent of the Legislative Yuan, appointed by the President of the Republic.

Article 105. The Auditor General shall within three months after presentation by the Executive Yuan of the final accounts of revenues and expenditures, complete the auditing thereof in accordance with law, and submit an auditing report to the Legislative Yuan.

Article 106. The organization of the Control Yuan shall be prescribed by law.

Chapter X. Powers of the Central and Local Governments

Article 107. In the following matters, the Central Government shall have the power of legislation and administration:
1. Foreign affairs;
2. National defense and military affairs concerning national defense;
3. Nationality law and criminal, civil and commercial law;
4. Judicial system;
5. Aviation, national highways, state-owned railways, navigation, postal and telegraph service;
6. Central Government finance and national revenues;
7. Demarcation of national, provincial and *hsien* revenues;
8. State-operated economic enterprises;
9. Currency system and state banks;
10. Weights and measures;
11. Foreign trade policies;
12. Financial and economic matters affecting foreigners or foreign countries; and
13. Other matters relating to the Central Government as provided by this Constitution.

Article 108. In the following matters, the Central Government shall have the power of legislation and administration, but the Central Government may delegate the power of administration to the provincial and *hsien* governments:

1. General Principles of Provincial and *Hsien* Self-Government;
2. Division of administrative areas;
3. Forestry, industry, mining and commerce;
4. Educational system;
5. Banking and exchange system;
6. Shipping and deep-sea fishery;
7. Public utilities;
8. Co-operative enterprises;
9. Water and land communication and transportation covering two or more provinces;
10. Water conservancy, waterways, agriculture and pastoral enterprises covering two or more provinces;
11. Registration, employment, supervision, and security of tenure of officials in central and local governments;
12. Land legislation;
13. Labor legislation and other social legislation;
14. Eminent domain;
15. Census-taking and compilation of population statistics for the whole country;
16. Immigration and land reclamation;
17. Police system;
18. Public health;
19. Relief, pecuniary aid in case of death and aid in case of unemployment; and
20. Preservation of ancient books and articles and sites of cultural value.

With respect to the various items enumerated in the preceding paragraph, the provinces may enact separate rules and regulations, provided these are not in conflict with national laws.

Article 109. In the following matters, the provinces shall have the power of legislation and administration, but the provinces may delegate the power of administration to the *hsien*;

1. Provincial education, public health, industries and communications;
2. Management and disposal of provincial property;

3. Administration of municipalities under provincial jurisdiction;

4. Province-operated enterprises;

5. Provincial cooperative enterprises;

6. Provincial agriculture, forestry, water conservancy, fishery, animal husbandry and public works;

7. Provincial finance and revenues;

8. Provincial debts;

9. Provincial banks;

10. Provincial police administration;

11. Provincial charitable and public welfare works; and

12. Other matters delegated to the provinces in accordance with national laws.

Except as otherwise provided by law, any of the matters enumerated in the various items of the preceding paragraph, in so far as it covers two or more provinces, may be undertaken jointly by the provinces concerned.

When any province, in undertaking matters listed in any of the items of the first paragraph, finds its funds insufficient, it may, by resolution of the Legislative Yuan, obtain subsidies from the National Treasury.

Article 110. In the following matters, the *hsien* shall have the power of legislation and administration:

1. *Hsien* education, public health, industries and communications;

2. Management and disposal of *hsien* property;

3. *Hsien*-operated enterprises;

4. *Hsien* cooperative enterprises;

5. *Hsien* agriculture and forestry, water conservancy, fishery, animal husbandry and public works;

6. *Hsien* finance and revenues;

7. *Hsien* debts;

8. *Hsien* banks;

9. Administration of *hsien* police and defense;

10. *Hsien* charitable and public welfare works; and

11. Other matters delegated to the *hsien* in accordance with national laws and Provincial Self-Government Regulations.

Except as otherwise provided by law, any of the matters enumerated in the various items of the preceding paragraph, in so far as it covers two or more *hsien*, may be undertaken jointly by the *hsien* concerned.

Article 111. Any matter not enumerated in Articles 107, 108, 109, and 110 shall fall within the jurisdiction of the Central Government, if it is national in nature; of the province, if it is provincial in nature; and of the *hsien*, if it concerns the *hsien*. In case of dispute, it shall be settled by the Legislative Yuan.

Chapter XI. System of Local Government

Section 1. The Province

Article 112. A province may convoke a Provincial Assembly to enact, in accordance with the General Principles of Provincial and *Hsien* Self-Government, regulations, provided the said regulations are not in conflict with the Constitution.

The organization of the Provincial Assembly and the election of the delegates shall be prescribed by law.

Article 113. The Provincial Self-Government Regulations shall include the following provisions:

1. In the province, there shall be a Provincial Council. Members of the Provincial Council shall be elected by the people of the province.

2. In the province, there shall be a Provincial Government with a Provincial Governor who shall be elected by the people of the province.

3. Relationship between the province and the *hsien*.

The legislative power of the province shall be exercised by the Provincial Council.

Article 114. The Provincial Self-Government Regulations shall, after enactment, be forthwith submitted to the Judicial Yuan. The Judicial Yuan, if it deems any part thereof unconstitutional, shall declare null and void the articles repugnant to the Constitution.

Article 115. If, during the enforcement of the Provincial Self-Government Regulations, there should arise any serious obstacle in the application of any of the articles contained therein, the Judicial Yuan shall first summon the various parties concerned to present their views; and thereupon the Presidents of the Executive Yuan, Legislative Yuan, Judicial Yuan, Examination Yuan and Control Yuan shall form a

Committee, with the President of the Judicial Yuan as Chairman, to propose a formula for solution.

Article 116. Provincial rules and regulations that are in conflict with national laws shall be null and void.

Article 117. When doubt arises as to whether or not there is a conflict between provincial rules or regulations and national laws, interpretation thereon shall be made by the Judicial Yuan.

Article 118. The self-government of municipalities under the direct jurisdiction of the Executive Yuan shall be prescribed by law.

Article 119. The local self-government system of the Mongolian Leagues and Banners shall be prescribed by law.

Article 120. The self-government system of Tibet shall be safeguarded.

Section 2. The *Hsien*

Article 121. The *hsien* shall enforce *hsien* self-government.

Article 122. A *hsien* may convoke a *hsien* assembly to enact, in accordance with the General Principles of Provincial and *Hsien* Self-Government, *Hsien* Self-Government Regulations, provided the said regulations are not in conflict with the Constitution or with Provincial Self-Government Regulations.

Article 123. The people of the *hsien* shall, in accordance with law, exercise the rights of initiative and referendum in matters within the sphere of *hsien* self-government, and shall, in accordance with law, exercise the rights of election and recall of the magistrate and other *hsien* self-government officials.

Article 124. In the *hsien*, there shall be a *hsien* council. Members of the *hsien* council shall be elected by the people of the *hsien*.
The legislative power of the *hsien* shall be exercised by the *hsien* council.

Article 125. *Hsien* rules and regulations that are in conflict with national laws, or with provincial rules and regulations, shall be null and void.

Article 126. In the *hsien*, there shall be a *hsien* government with a *hsien* magistrate who shall be elected by the people of the *hsien*.

Article 127. The *hsien* magistrate shall have charge of *hsien* self-government and shall administer matters delegated to the *hsien* by the central or provincial government.

Article 128. The provisions governing the *hsien* shall apply *mutatis mutandis* to the municipality.

Chapter XII. Election, Recall, Initiative and Referendum

Article 129. The various kinds of elections prescribed in this Constitution, except as otherwise provided by this Constitution, shall be by universal, equal, and direct suffrage and by secret ballot.

Article 130. Any citizen of the Republic of China who has attained the age of 20 years shall have the right of election in accordance with law. Except as otherwise provided by this Constitution or by law, any citizen who has attained the age of 23 years shall have the right of being elected in accordance with law.

Article 131. All candidates in the various kinds of elections prescribed in this Constitution shall openly campaign for their election.

Article 132. Intimidation or inducement shall be strictly forbidden in elections. Suits arising in connection with elections shall be tried by the courts.

Article 133. A person elected may, in accordance with law, be recalled by his constituency.

Article 134. In the various kinds of elections, the number of women to be elected shall be fixed, and measures pertaining thereto shall be prescribed by law.

Article 135. The number of delegates to the National Assembly and the manner of their election from people in interior areas, who have their own conditions of living and habits, shall be prescribed by law.

Article 136. The exercise of the rights of initiative and referendum shall be prescribed by law.

Chapter XIII. Fundamental National Policies

Section 1. National Defense

Article 137. The national defense of the Republic of China shall have as its objective the safeguarding of national security and the preservation of world peace.

The organization of national defense shall be prescribed by law.

Article 138. The land, sea and air forces of the whole country shall be above personal, regional, or party affiliations, shall be loyal to the state and shall protect the people.

Article 139. No political party and no individual shall make use of armed forces as an instrument in a struggle for political powers.

Article 140. No military man in active service may concurrently hold a civil office.

Section 2. Foreign Policy

Article 141. The foreign policy of the Republic of China shall, in a spirit of independence and initiative and on the basis of the principles of equality and reciprocity, cultivate good-neighborliness with other nations, and respect treaties and the Charter of the United Nations, in order to protect the rights and interests of Chinese citizens residing abroad, promote international co-operation, advance international justice and ensure world peace.

Section 3. National Economy

Article 142. National economy shall be based on the Principle of the People's Livelihood and shall seek to effect equalization of land ownership and restriction of private capital in order to attain a well-balanced sufficiency in national wealth and people's livelihood.

Article 143. All land within the territory of the Republic of China shall belong to the whole body of citizens. Private ownership of land, acquired by the people in accordance with law, shall be protected and restricted by law. Privately-owned land shall be liable to taxation according to its value, and the Government may buy such land according to its value.

Mineral deposits which are embedded in the land, and natural power which may, for economic purposes, be utilized for the public benefit shall belong to the state, regardless of the fact that private individuals may have acquired ownership over such land.

If the value of a piece of land has increased, not through the exertion of labor or the employment of capital, the State shall levy thereon an increment tax, the proceeds of which shall be enjoyed by the people in common.

In the distribution and readjustment of land, the State shall, in principle, assist self-farming land-owners and persons who make use of the land by themselves, and shall also regulate their appropriate areas of operation.

Article 144. Public utilities and other enterprises of a monopolistic nature shall, in principle, be under public operation. In cases permitted by law, they may be operated by private citizens.

Article 145. With respect to private wealth and privately-operated enterprises, the State shall restrict them by law if they are deemed detrimental to a balanced development of national wealth and people's livelihood.

Cooperative enterprises shall receive encouragement and assistance from the state.

Private citizens' productive enterprises and foreign trade shall receive encouragement, guidance and protection from the State.

Article 146. The State shall, by the use of scientific techniques, develop water conservancy, increase the productivity of land, improve agricultural conditions, plan for the utilization of land, develop agricultural resources and hasten the industrialization of agriculture.

Article 147. The Central Government, in order to attain a balanced economic development among the provinces, shall give appropriate aid to poor or unproductive provinces.

The provinces, in order to attain a balanced economic development among the *hsien*, shall give appropriate aid to poor or unproductive *hsien*.

Article 148. Within the territory of the Republic of China, all goods shall be permitted to move freely from place to place.

Article 149. Financial institutions shall, in accordance with law, be subject to State control.

Article 150. The State shall extensively establish financial institutions for the common people, with a view to relieving unemployment.

Article 151. With respect to Chinese citizens residing abroad, the State shall foster and protect the development of their economic enterprises.

Section 4. Social Security

Article 152. The State shall provide suitable opportunity for work to people who are able to work.

Article 153. The State, in order to improve the livelihood of labourers and farmers and to improve their productive skill, shall enact laws and carry out policies for their protection.

Women and children engaged in labour shall, according to their age and physical condition, be accorded special protection.

Article 154. Capital and labor shall, in accordance with the principle of harmony and cooperation, promote productive enterprises. Conciliation and arbitration of disputes between capital and labor shall be prescribed by law.

Article 155. The state, in order to promote social welfare, shall establish a social insurance system. To the aged and the infirm who are unable to earn a living, and to victims of unusual calamities, the State shall give appropriate assistance and relief.

Article 156. The State, in order to consolidate the foundation of national existence and development, shall protect motherhood and carry out the policy of promoting the welfare of women and children.

Article 157. The State, in order to improve national health, shall establish extensive services for sanitation and health protection, and a system of public medical service.

Section 5. Education and Culture

Article 158. Education and culture shall aim at the development among the citizens of the national spirit, the spirit of self-government, national morality, good physique, scientific knowledge and the ability to earn a living.

Article 159. All citizens shall have equal opportunity to receive an education.

Article 160. All children of school age from 6 to 12 years shall receive free primary education. Those from poor families shall be supplied with books by the Government.

All citizens above school age who have not received primary education shall receive supplementary education free of charge and shall also be supplied with books by the Government.

Article 161. The national, provincial, and local governments shall extensively establish scholarships to assist students of good scholastic standing and exemplary conduct who lack the means to continue their school education.

Article 162. All public and private educational and cultural institutions in the country shall, in accordance with law, be subject to State supervision.

Article 163. The State shall pay due attention to the balanced development of education in different regions, and shall promote social education in order to raise the cultural standard of the citizens in general. Grants from the National Treasury shall be made to frontier regions and economically poor areas to help them meet their educational and cultural expenses. The Central Government may either itself undertake the more important educational and cultural enterprises in such regions or give them financial assistance.

Article 164. Expenditures of educational programs, scientific studies and cultural services shall not be, in respect of the Central Government, less than 15 per cent of the total national budget; in respect of each province, less than 25 per cent of the total provincial budgets; and in respect of each municipality or *hsien*, less than 35 per cent of the total municipal or *hsien* budget. Educational and cultural foundations established in accordance with law shall, together with their property, be protected.

Article 165. The State shall safeguard the livelihood of those who work in the fields of education, sciences, and arts, and shall, in accordance with the development of national economy, increase their remuneration from time to time.

Article 166. The State shall encourage scientific discoveries and inventions, and shall protect ancient sites and articles of historical, cultural or artistic value.

Article 167. The State shall give encouragement or subsidies to the following enterprises or individuals;

1. Educational enterprises in the country which have been operated with good record by private individuals;

2. Educational enterprises which have been operated with good record by Chinese citizens residing abroad;

3. Persons who have made discoveries or inventions in the fields of learning and technology; and

4. Persons who have rendered long and meritorious services in the field of education.

Section 6. Frontier Regions

Article 168. The State shall accord to the various racial groups in the frontier regions legal protection of their status and shall give them special assistance in their local self-government undertakings.

Article 169. The State shall, in a positive manner, undertake and foster the development of education, culture, communications, water conservancy, public health and other economic and social enterprises of the various racial groups in the frontier regions. With respect to the utilization of land, the State shall, after taking into account the climatic conditions, the nature of the soil and the life and habits of the people, adopt measures to protect the land and to assist in its development.

Chapter XIV. Enforcement and Amendment of the Constitution

Article 170. The term 'law,' as used in this Constitution, shall denote any legislative bill that shall have been passed by the Legislative Yuan and promulgated by the President of the Republic.

Article 171. Laws that are in conflict with the Constitution shall be null and void.

When doubt arises as to whether or not a law is in conflict with the Constitution, interpretation thereon shall be made by the Judicial Yuan.

Article 172. Ordinances that are in conflict with the Constitution or with laws shall be null and void.

Article 173. The Constitution shall be interpreted by the Judicial Yuan.

Article 174. Amendments to the Constitution shall be made in accordance with one of the following procedure:

1. Upon the proposal of one-fifth of the total number of the delegates to the National Assembly and by a resolution of three-fourths of the delegates present at a meeting having a quorum of two-thirds of the entire Assembly, the Constitution may be amended.

2. Upon the proposal of one-fourth of the members of the Legislative Yuan and by a resolution of three-fourths of the members present at a meeting having a quorum of three-fourths of the members of the Yuan, an amendment may be drawn up and submitted to the National Assembly by way of referendum. Such a proposed amendment to the Constitution shall be publicly published half a year before the National Assembly convenes.

Article 175. Whenever necessary, enforcement procedures in regard to any matters prescribed in this Constitution shall be separately provided by law.

The preparatory procedures for the enforcement of this Constitution shall be decided upon by the same National Assembly which shall have adopted this Constitution.

Additional Articles of the Constitution of the Republic of China

(Articles 1 through 10 were adopted by the second extraordinary session of the First National Assembly at its sixth plenary meeting on 22 April 1991, and promulgated by the President on 1 May 1991. Articles 11 through 18 were adopted by the extraordinary session of the Second National Assembly at its 27th plenary meeting on 27 May 1992, and promulgated by the President on 28 May 1992.)

To meet the requisites of national unification, the following additional articles are added to the ROC Constitution in accordance with Article 27, Paragraph 1, Item 3, and Article 174, Item 1:

Article 1. Members of the National Assembly shall be elected according to the following provisions without being subject to the restrictions in Articles 26 and 135 of the Constitution:

1. Two members shall be elected from each Special Municipality, each county or city in the free area. However, where the population exceeds 100,000 persons, one member shall be added for each additional 100,000 persons.

2. Three members each shall be elected from the lowland and highland aborigines in the free area.

3. Twenty members shall be elected from the Chinese citizens who reside abroad.

4. Eighty members shall be elected from one nationwide constituency.

If the number of seats allotted to a Special Municipality, county or city covered under Item One (1) above; or if the number of seats won by a political party under Item Three (3) or Four (4) above is between five and ten, then one of the seats stipulated in the pertaining item shall be reserved for a female candidate. Where the number exceeds ten, one seat out of each additional ten shall be reserved for a female candidate.

Article 2. Members of the Legislative Yuan shall be elected according to the following provisions without being subject to the restrictions in Article 64 of the Constitution:

1. Two members shall be elected from each province and each Special Municipality in the free area. Where the population exceeds 200,000 persons, however, one member will be added for each additional 100,000 persons; and where the population exceeds one million persons, one member will be added for each additional 200,000 persons.

2. Three members each shall be elected from the lowland and highland aborigines in the free area.

3. Six members shall be elected from the Chinese citizens who reside abroad.

4. Thirty members shall be elected from one nationwide constituency.

If the number of seats allotted to a province or Special Municipality covered under Item One (1) above; or if the number of seats won by a political party under Item Three (3) or Four (4) above is between five and ten, then one of the seats stipulated in the pertaining item shall be reserved for a female candidate. Where the number exceeds ten, one seat out of each additional ten shall be reserved for a female candidate.

Article 3. Members of the Control Yuan shall be elected by the Provincial Assembly and municipal councils according to the following provisions without being subject to the restrictions in Article 91 of the Constitution:

1. Twenty-five members shall be elected from Taiwan Province in the free area.

2. Ten members shall be elected from each Special Municipality in the free area.

3. Two members shall be elected from the Chinese citizens who reside abroad.

4. Five members shall be elected from one nationwide constituency.

If the number of seats allotted to Taiwan Province or from a Special Municipality covered under Item One (1) or Two (2) above; or if the number of seats won by a political party under Item Four (4) above is between five and ten, then one of the seats stipulated in the pertaining item shall be reserved for a woman. Where the number of seats exceeds ten, one seat out of each additional ten shall be reserved for a woman.

The number of Provincial Assembly members who can be elected to the Control Yuan is limited to two; the number of members from each municipal council who can be elected to the Control Yuan is limited to one.

Article 4. The election of members of the National Assembly, Legislative Yuan, and Control Yuan, or recall of any thereof, shall be conducted in accordance with the provisions contained in the Public Officials Election and Recall Law. The members representing Chinese citizens residing abroad and the members representing the nationwide constituency shall be elected by way of party-list proportional representation.

Article 5. Members of the Second National Assembly shall be elected before 31 December 1991. Their term of office begins on 1 January 1992 and expires on the day when the members of the Third National Assembly meet, pursuant to Article 29 of the Constitution prior to the expiration of the 8th presidential term in 1996. This is not subject to restrictions imposed by Paragraph 1 of Article 28 of the Constitution.

Those additional members of the National Assembly elected in Taiwan pursuant to the Temporary Provisions effective during the Period of National Mobilization for Suppression of the Communist Rebellion shall exercise their official powers together with the members of the Second National Assembly until 31 January 1993.

The Second-Term Legislative Yuan members and the Second-Term Control Yuan Members shall be elected prior to 31 January 1993, and shall begin to exercise their official powers on 1 February 1993.

Article 6. An extraordinary session of the National Assembly shall be convened by the President within three months after the members of the Second National Assembly are elected so that the National Assembly may exercise the powers granted by Article 27, Paragraph 1, Item 3 of the Constitution.

Article 7. The President may, by resolution of a council of the Executive Yuan, issue emergency orders and take all necessary measures to avert an imminent danger to the security of the State or of the people or to cope with any serious financial or economic crisis, without being subject to the restrictions prescribed in Article 43 or the Constitution. However, such orders shall, within 10 days of issuance, be presented to the Legislative Yuan for confirmation. Should the Legislative Yuan withhold confirmation, the said emergency orders shall forthwith cease to be valid.

Article 8. If any laws originally intended to be applicable solely during the Period of National Mobilization for Suppression of the Communist Rebellion shall not have been revised by the termination of the Period of National Mobilization for Suppression of the Communist Rebellion, said laws shall remain in effect until 31 July 1992.

Article 9. To determine major policies for national security, the President may establish the National Security Council and its subsidiary organ, the National Security Bureau.

The Executive Yuan may establish the Central Personnel Administration.

The organizations of the above two paragraphs shall be established according to law. Before the legislative process is completed, the current organizational statutes shall remain in force till 31 December 1993.

Article 10. Rights and obligations between the people of the mainland China area and those of the free area, and the disposition of other related affairs shall be specially regulated by law.

Article 11. In addition to the exercise of its powers and obligations pursuant to Article 27 of the Constitution, the National Assembly shall also exercise its right to confirm the appointment of personnel nominated by the President in accordance with Additional Article 13, Paragraph 1; Additional Article 14, Paragraph 2; and Additional Article 15, Paragraph 2.

The aforementioned right of confirmation shall be exercised at an extraordinary session of the National Assembly convoked by the President and shall not be subject to the restrictions in Article 30 of the Constitution.

When the National Assembly convenes, it shall hear a report on the state of the nation by the President, discuss national affairs, and offer counsel. In the event that the National Assembly has not convened for over a year, the President shall convoke an extraordinary session for the aforementioned purpose notwithstanding the restrictions in Article 30 of the Constitution.

Beginning with the Third National Assembly, delegates to the National Assembly shall be elected every four years and the provisions in Article 28, Paragraph 1 of the Constitution shall not apply.

Article 12. Effective from the 1996 election for the ninth-term President and Vice-President, the President and the Vice-President shall be elected by the entire electorate in the free area of the Republic of China.

The electoral method for the aforementioned election shall be formulated in the Additional Articles to the Constitution at an extraordinary session of the National Assembly to be convoked by the President before 20 May 1995.

Beginning with the ninth presidential term, the term of office for both the President and the Vice-President shall be four years. The President and the Vice-President may be re-elected for a second term; and the provisions in Article 47 of the Constitution shall not apply.

Recall of the President and the Vice-President shall be executed in accordance with the following provisions:

1. By a motion to recall put forward by one-fourth of all delegates to the National Assembly, and passed with the concurrence of two-thirds of such delegates.

2. By a resolution to impeach adopted by the Control Yuan, and passed as a resolution to recall by two-thirds of all delegates to the National Assembly.

Should the office of the Vice-President become vacant, the President shall nominate a candidate within three months and convoke an extraordinary session of the National Assembly to elect a new Vice-President, who shall serve out the original term until its expiration. Should the offices of both the President and the Vice-President become vacant, the president of the Legislative Yuan shall serve notice on the National Assembly to convoke an extraordinary session within three months to elect a new President and a new Vice-President, who shall serve out each respective original term until its expiration.

Article 13. The Judicial Yuan shall have a President, a Vice-President, and a certain number of Grand Justices, all of whom shall be nominated and, with the consent of the National Assembly, appointed by the President; and the pertinent provisions in Article 79 of the Constitution shall not apply.

The Grand Justices of the Judicial Yuan shall, in addition to discharging their duties according to Article 78 of the Constitution, also form a Constitutional Tribunal to adjudicate matters relating to the dissolution of unconstitutional political parties.

A political party shall be unconstitutional if its goals or activities jeopardize the existence of the Republic of China or free, democratic constitutional order.

Article 14. The Examination Yuan shall be the highest examination body of the State, and shall be responsible for the following matters; and the provisions in Article 83 of the Constitution shall not apply to:

1. All examination-related matters;

2. All matters relating to the qualification screening, security of tenure, pecuniary aid in case of death, and retirement of civil servants; and

3. All legal matters relating to the employment, discharge, performance evaluation, scale of salaries, promotion, transfer, commendation and award for civil servants.

The Examination Yuan shall have a President, a Vice-President, and several members, all of whom shall be nominated and, with the consent of the National Assembly, appointed by the President; and the provisions in Article 84 of the Constitution shall not apply.

The provisions in Article 85 of the Constitution concerning holding examinations in different areas, with prescribed numbers of persons to be selected according to various provinces and areas, shall cease to apply.

Article 15. The Control Yuan shall be the highest control body of the State and shall exercise the powers of impeachment, censure and audit; and the provisions in Articles 90 and 94 of the Constitution concerning exercising the power of consent shall not apply.

The Control Yuan shall have 29 members, including a President and a Vice-President, all of whom shall serve a term of six years and shall be nominated and, with the consent of the National Assembly, appointed by the President. The provisions in Article 91 through 93, and in Additional Articles 3 and 4, as well as Article 5, Paragraph 3 of the Constitution concerning the members of the Control Yuan shall cease to be applicable.

Impeachment proceedings by the Control Yuan against a public functionary in the Central Government, any local government, or against personnel of the Judicial Yuan or the Examination Yuan shall be initiated by two or more members of the Control Yuan, and be investigated and voted upon by a committee of not less than nine of its members notwithstanding the restrictions in Article 98 of the Constitution.

In the case of impeachment by the Control Yuan of Control Yuan personnel for dereliction of duty or violation of the law, the provisions of Article 95 and Article 97, Paragraph 2 of the Constitution, as well as the foregoing paragraph shall apply.

A motion by the Control Yuan impeaching the President or the Vice-President must be initiated by more than half of all the members of the Control Yuan and passed by more than two-thirds of all such members for it to be submitted to the National Assembly notwithstanding the restrictions in Article 100 of the Constitution.

Members of the Control Yuan must be beyond party affiliation and independently exercise their powers and discharge their responsibilities in accordance with the law.

The provisions in Articles 101 and 102 of the Constitution shall cease to apply.

Article 16. Provisions of Additional Article 15, Paragraph 2 shall take effect with the nomination of Second-Term Control Yuan members.

The Second-Term Control Yuan members shall assume their offices on 1 February 1993. Provisions of Additional Article 15, Paragraph 1, and Paragraphs 3 through 7 shall take effect on the same date.

Provisions of Additional Article 13, Paragraph 1 and Article 14, Paragraph 2 relating to the appointment of the personnel of the Judicial Yuan and the Examination Yuan shall take effect on 1 February 1993. Nominations of personnel made before 31 January 1993 shall still be approved by the Control Yuan before appointment by the President. Incumbent personnel, however, need not be re-nominated and re-appointed before the expiration of their terms.

Article 17. The system of local governments in the provinces and counties shall include the following provisions, which shall be established by the enactment of appropriate laws notwithstanding the restrictions in Article 108, Paragraph 1, Item 1; Article 112 through 115; and Article 122 of the Constitution:

1. There shall be a provincial assembly in each province and a county council in each county. Members of the provincial assembly and the county council shall be elected by the people of the province and the people of the county, respectively.

2. The legislative power of a province and that of a county shall be exercised by the provincial assembly and the county council, respectively.

3. In a province, there shall be a provincial government with a provincial governor. In a county, there shall be a county government with a county magistrate. The provincial governor and the county magistrate shall be elected by the people of the province and the people of the county, respectively.

4. The relationship between the province and the county.

5. The self-governance of provinces is subject to supervision by the Executive Yuan, while the self-governance of counties is subject to supervision by the provincial government.

Article 18. The State shall encourage development of and investment in science and technology, facilitate the upgrade of industry, promote the modernization of agriculture and fishery, emphasize the exploitation and utilization of water resources, and intensify international economic cooperation.

Environmental and ecological protection shall be given equal consideration with economic and technological development.

The State shall inaugurate universal health insurance coverage and promote the research and development of both modern and traditional medicines.

The State shall protect the dignity of women, safeguard their personal safety, eliminate sexual discrimination, and further substantive equality between the sexes.

The State shall safeguard the rights of the handicapped and disabled to insurance, medical care, education, training, employment assistance, support for daily living needs and relief, so as to help them attain independence and further their careers.

The State shall accord to the aborigines in the free area legal protection of their status and the right to political participation. It shall also provide assistance and encouragement for their education, cultural preservation, social welfare and business undertakings. The same protection and assistance shall be given to the people of Kinmen˘ and Matsu areas.

The State shall accord to Chinese nationals residing overseas protection of their rights to political participation.

(Translation provided by the Government Information Office, July 1992.)

Bibliography

I. Publications with Focus on Taiwan

Ahern, Emily Martin and Gates, Hill (eds.), *The Anthropology of Taiwanese Society*. Stanford: Stanford University Press, 1981.

Annual Review of Government Administration, ROC.

Appleton, Sheldon, 'The Social and Political Impact of Education in Taiwan,' *Asian Survey*, 16 (August 1976): 703–20.

———, 'Taiwanese and Mainlanders on Taiwan: A Survey of Student Attitudes,' *China Quarterly*, 44 (October–December 1970): 38–65.

———, 'Survey Research On Taiwan,' *Public Opinion Quarterly*, 40 (Winter 1977): 468–81.

———, 'The Political Socialization of Taiwan's College Students,' *Asian Survey*, 10 (October 1970): 910–23.

Asian Wall Street Journal (AWSJ)

Ballantine, Joseph W., *Formosa, a Problem for United States Foreign Policy*. Washington, DC: Brookings Institution, 1952.

Baloyra, Enrique A. (ed.), *Comparing New Democracies*. Boulder: Westview Press, 1987.

Bate, H. MacLear, *Report from Formosa*. New York: E. P. Dutton, 1952.

Bessac, Frank B., *An Example of Social Change in Taiwan Related to Land Reform*. Missoula: University of Montana, 1967.

Bianco, Lucien, *Origins of the Chinese Revolution, 1915–1949*. Stanford: Stanford University Press, 1971.

Central Daily News, a government newspaper published in Taipei.

Central Election Commission (CEC), *Record of 1986 Election on Central Representatives in Free Areas During the Emergency Period*. Taipei: CEC, 1987a (in Chinese).

———, *The Voting Behaviors of the Voters*. Taipei: CEC, 1987b (in Chinese).

———, *Introduction to the Election in the Republic of China, Statistics for 1985 & 1986 Local Elections*. Taipei: CEC, 1986a (in Chinese).

———, *Record of 1985 Provincial and City Assembly Election*. Taipei: CEC, 1986b (in Chinese).

———, *General Conditions of the Elections in the Republic of China*. Taipei: CEC, 1984a (in Chinese).

———, *Introduction to the Election in the Republic of China, Record of 1983 Legislative Yuan Election*. Taipei: CEC, 1984b (in Chinese).

———, *Election Statistics of the Republic of China (1946–82)*. Taipei: CEC, 1982 (in Chinese).

Chai, Trong R., 'The Future of Taiwan,' *Asian Survey*, 26 (December 1986): 1309–23.

Chang, Hao, 'Change of KMT: from Revolutionary Religion to Reformism,' *China Times*, 28 March 1988: 3.

Chang, Hsiao-tsun; Hsiao, Hsin-huan; and Hsu, Chen-kuang (eds.), *Social Transformation, 1985 Taiwan Social Critique*. Taipei: Dun-Li Press, 1986 (in Chinese).

Chang, Mao-kuei, *Social Movement and Political Transformation*. Taipei: Institute for National Policy Research, 1989 (in Chinese).

Chang, Yu-mei, *The Min-Chuan Thought of the National Father*. Taipei: Yiu-shih, 1969 (in Chinese).

Chao, Yung-mau, 'An Analysis on Local Faction Participation and Democratic Value Orientation,' *The Annals, Chinese Association of Political Science*, 14 (December 1986): 59–127.

Chen, Hao and Kou, Wei-yong, *Examining the Factions in the KMT*. Taipei: Fongyun, 1985 (in Chinese).

Chen, Lung-chu and Lasswell, Harold D., *Formosa, China, and the United Nations*. New York: St. Martins Press, 1967.

Cheng, Tung-jen, 'Democratizing the Quasi-Leninist Regime in Taiwan,' *World Politics*, 41 (July 1989): 471–99.

Cheng, Tung-jen, and Haggard, Stephan (eds.), *Political Change in Taiwan*. Boulder: Lynne Rienner Publishers, 1992.

China Times, a daily newspaper published in Taiwan.

Chiou, C. L., 'Politics of Alienation and Polarization: Taiwan's "Tangwai" in the 1980s,' *Bulletin of Concerned Asian Scholars*, 18 (July/September 1986): 16–28.

Chiu, Hai-yuan and Chang, Yinh-wa (eds.), *Social and Cultural Change in Taiwan*. 2 vols. Taipei: Academia Sinica, 1986 (in Chinese).

Chiu, Hungdah (ed.), *China and the Taiwan Issue*. New York: Praeger Publishers, 1979.

Chou, Tsu-cheng, 'Electoral Competition and the Development of Opposition in Taiwan,' *The Annals* (Chinese Association of Political Science), 20 (December 1992): 75–120 (in Chinese).

Chou, Yangsun and Nathan, Andrew J., 'Democratizing Transition in Taiwan,' *Asian Survey*, 27 (March 1987): 277–99.

Chou, Yu-kou, *The First Thousand Days of Lee Teng-hui*. Taipei: Mai-tien Press, 1993 (in Chinese).

Chu, Yun-han, *Crafting Democracy in Taiwan*. Taipei: Institute for National Policy Research, 1992.

Clark, Cal, 'The Taiwan Exception: Implications for Contending Political Economy Paradigms,' *International Studies Quarterly*, 31 (March 1987): 327–56.

Cohen, Marc J., 'One China or Two: Facing Up to the Taiwan Question,' *World Policy Journal*, 4 (Fall 1987): 621–49.

Cohen, Myron L., *House United, House Divided: The Chinese Family in Taiwan.* New York: Columbia University Press, 1976.

Copper, John F., 'The KMT's 13th Party Congress: Reform, Renovation, New Blood, New Policies,' paper presented at the Sixth Annual Conference on Asian Topics, St. John's University, New York, 1988.

———, *Taiwan: Nation-State or Province?* Boulder: Westview Press, 1990.

———, 'Taiwan's 1992 Legislative Yuan Election,' *World Affairs*, 155 (Fall 1992): 71–9.

———, 'The Role of Minor Political Parties in Taiwan,' *World Affairs*, 155 (Winter 1993): 95–108.

———, *A Quiet Revolution: Political Development in the Republic of China.* Washington, DC: Ethics and Public Policy Center, 1988.

Copper, John F. and Chen, George P., *Taiwan's Elections: Political Development and Democratization in the Republic of China.* Baltimore: School of Law University of Maryland, Occasional Papers/Reprints Series in Contemporary Asian Studies, 1984.

Council for Economic Planning and Development, *Taiwan Statistical Data Book.* Taipei: Council for Economic Planning and Development, ROC, 1992.

Croizier, Ralph C., *Koxinga and Chinese Nationalism, History, Myth, and the Hero.* Cambridge, MA: East Asian Research Center, Harvard University, 1977.

Davidson, D. W. S., 'Politics of the Left in Taiwan,' *Bulletin of Concerned Asian Scholars*, 12 (April/June 1980): 18–24.

Deng, Pei-yun, *Taiwanese Student Movements in the Eighties.* Taipei: Chien-wei Press, 1993 (in Chinese).

Directorate-General of Budget, Accounting, and Statistics (BAS), Executive Yuan, *Statistical Abstract of National Income in Taiwan Area, Republic of China 1951–87.* Taipei: Directorate-General of BAS, 1988.

———, *Quarterly National Economic Trends, Taiwan Area, the Republic of China*, 64 (February 1994).

———, *Statistical Abstract of National Income in Taiwan Area, Republic of China 1951–1992.* Taipei: Directorate-General of BAS, 1993.

———, *National Income in Taiwan Area of the Republic of China 1992.* Taipei: Directorate-General of BAS, 1992.

———, *Report on the Survey of Personal Income Distribution in Taiwan Area, Republic of China, 1986.* Taipei: Directorate-General of BAS, 1987a.

———, *Social Indicators of the Republic of China, 1986.* Taipei: Directorate-General of BAS, 1987b (in Chinese).

———, *Social Indicators of the Republic of China, 1991*. Taipei: Directorate-General of BAS, 1991 (in Chinese).

Domes, Jurgen, 'Political Differentiation in Taiwan's Group Formation within the Ruling Party and the Opposition Circles, 1979–1980,' *Asian Survey*, 21 (October 1981): 1011–28.

———, 'Taiwan in 1992: On the Verge of Democracy,' *Asian Survey*, 33 (January 1993): 54–60.

Dreyer, June Teuful, 'Taiwan's December 1991 Election,' *World Affairs*, 155 (Fall 1992): 67–70.

Du, Nianchong and Yang, Chun-se (eds.), *Confucius Ethics and Economic Development*. Taipei: Yun-chen Cultural Enterprise, 1988 (in Chinese).

Eberstadt, Nicholas, 'Taiwan and South Korea: The "Democratization" of Outlier State,' *World Affairs*, 155 (Fall 1992): 80–9.

The Economist.

Far Eastern Economic Review (*FEER*).

Fongyun Press, *Examining the Non-partisan Forces*. 2nd edn. Taipei: Fong-yun Press, 1985 (in Chinese).

Francis, Corinna-Barbara, 'Interview with Kang Ning-hsiang,' *Bulletin of Concerned Asian Scholars*, 17 (July/September 1985): 57–64.

Gold, Thomas B., *State and Society in the Taiwan Miracle*. Armonk, NY: M. E. Sharpe, Inc., 1986.

Ho, Szu-yin, *Open Registration for Political Parties and Its Impact on the 1989 Election*. Taipei: Commission of Research, Development, and Evaluation, Executive Yuan, 1991 (in Chinese).

———, *Legislative Politics of the Republic of China*. Ph.D. Dissertation, University of California, Santa Barbara, 1986.

Hsiao, Michael Hsin-huan (ed.), *Middle Class in the Transforming Taiwanese Society*. Taipei: Chu-liu Press, 1990 (in Chinese).

Hsu, Chen-kwang and Sung, Wen-li (eds.), *New Social Movements in Taiwan*. 3rd edn. Taipei: Chu-liu Press, 1992 (in Chinese).

Hu, Fo (ed.), *Voting Behavior and Election Culture*. Taipei: Chinese Political Science Association, 1986 (in Chinese).

Hu, Li-tai (ed.), *Cannot, and Will Not: Immortal Speeches in 1986*. Taipei: Jeao-da Cultural Press, 1987 (in Chinese).

Huang, Shiow-duan, 'To Serve the People: An Analysis of Constituency Service of the Supplementary Legislators,' *The Annals* (Chinese Association of Political Science), 20 (December 1992): 141–213 (in Chinese).

Huchang, Tongfu, *The Future of Taiwan, Interviews on the Reformers*. Taipei: Dong-tsa Press, 1988 (in Chinese).

Hwang, Teh-fu, 'Analysis on Collective Political Protest,' 1988a, a paper for the Collective Protest Conference in 1988 (in Chinese).

———, 'Response to Collective Protest: Analysis and Evaluation,' 1988b, a paper for the Collective Protest Conference in 1988 (in Chinese).

———, *The Democratic Progressive Party and Democratization in Taiwan.* Taipei: Shih-yin Press, 1992 (in Chinese).

———, 'Local Factions, Party Competition, and Political Democratization in Taiwan,' *National Chengchi University Journal,* 61 (June 1990): 723–45.

———, 'Electoral Competition and Democratic Transition in the Republic of China,' *Issues and Studies,* 27 (October 1991): 97–123.

Independence Evening Post and *Independence Morning Post,* two private newspapers published in Taiwan.

The Journalist (Shin Shin Wen), a news weekly published in Taiwan.

Ju, Gao-jen; Liu, Ming-sen; Chen Fuwei (eds.), *A Lonely Opposition Philosopher-Ju Gao-jen.* Taipei: Ju Gaojen, 1988 (in Chinese).

Kagan, Richard C., 'Martial Law in Taiwan,' *Bulletin of Concerned Asian Scholars,* 14 (July/September 1982): 48–54.

Kang, Ning-hsing, *Six Years in Politics.* Taipei: Chang-chou Press, 1978.

———, *The Crisis and the Hope.* Taipei: The Eighties, 1973 (in Chinese).

Kao, Yin-mao, Chiu, Hungda (eds.), *Current Reform Issues of the Republic of China.* Taipei: China Times, 1988 (in Chinese).

Kerr, George H., *Formosa Betrayed.* Boston: Houghton Mifflin Company, 1965.

———, *Formosa, Lisensed Revolution and the Home Rule Movement, 1895–1945.* Honolulu: The University Press of Hawaii, 1974.

Kuo, Peter, 'Taiwan's News Media: Its Role in Democratization,' *World Affairs,* 155 (Winter 1993): 109–18.

Kuo, Shirley W. Y., *The Taiwan Economy in Transition,* Boulder: Westview Press, 1983.

Kuo, Shirley W. Y.; Ranis, Gustav; Fei, John C. H., *The Taiwan Success Story: Rapid Growth with Improved Distribution in the Republic of China, 1952–1979.* Boulder: Westview Press, 1981.

Lee, Yuan-chuan, 'Peasant Movements and Problems of the Agricultural Economy in Taiwan,' paper presented at the 1989 annual meeting of the Association for Asian Studies, Washington, DC.

Lei, Fei-long (ed.), *Voting Behavior in Transitional Society: Integrated Research on the Voters in Taiwan Area.* Taipei: National Chengchi University, 1986 (in Chinese).

——— (ed.), *Voting Behavior of Taiwan Electorate: Searching for a Theoretical Model.* Taipei: National Science Commission, Executive Yuan, 1991 (in Chinese).

Lerman, Arthur J., 'National Élite and Local Politician in Taiwan,' *American Political Science Review,* 71 (1977): 1407–22.

Li, Ao, *A Study on the February 28 Incident.* 3rd edn. Taipei: Lee Ao Press, 1991 (in Chinese).

Li, Kwoh-ting and Yu, Tzong-shian (eds.), *Experiences and Lessons of Economic Development in Taiwan.* Taipei: Academia Sinica, 1982.

Li, Ting-yi, *Contemporary History of China.* Taipei: Chonghua Book Store, 1978 (in Chinese).

Liberty Times, a daily newspaper published in Taiwan.

Lin, Bih-jaw (ed.), *Contemporary China and the Changing International Community.* Taipei: Institute of International Relations, 1993.

Lin, Chen-chieh (translator), *Green Utopia, Platform of West Germany's Green Party.* Taipei: Chien-gin Press, 1988 (in Chinese).

Lin, Chong-pin and Chan, Man-jung Mignon, 'Taiwan and Mainland China: A Comparison on Democratization,' *World Affairs,* 155 (Winter 1993): 117–29.

Lin, Mu-shun, *Taiwan's February Revolution.* Taipei: Chien Wei Press, 1992 (in Chinese).

Liu, Chin-ching, 'The Development of the NICs and the New Economic Status, the Political and Economic Bases of Democratization,' *May Forum,* 1 (May 1988): 3–32 (in Chinese).

Liu, I-chou, *The Electoral Effect of Social Context Control on Voters: The Case of Taipei, Taiwan.* Ph.D. Dissertation, The University of Michigan, 1990.

Long, Chong-ten, *Chiang Chin-kuo, Before and After.* Taipei: Shinmei Press, 1988 (in Chinese).

Lu, Hsiu-yi, 'The Opposition and Political Change in Taiwan,' paper presented at the 1989 annual meeting of the Association for Asian Studies, Washington, DC.

Ma, Chi-hua, 'The Principles of People and the Modern Life,' *Political Culture,* 2 (September 1985): 67–87 (in Chinese).

Mancall, Mark, *Formosa Today.* New York: Frederick A. Praeger Publisher, 1964.

McBeat, Gerald A., 'Youth Change in Taiwan, 1975–1985,' *Asian Survey,* 26 (September 1986): 1020–36.

Myers, Ramon H., 'Political Theory and Recent Political Developments in the Republic of China,' *Asian Survey,* 27 (September 1987): 1003–22.

Nan-min, 'Taiwanese Politics after Chiang Chin-kuo: the Trend of New Feudalism,' *Taiwan and the World,* 31 (May 1986): 17–25 (in Chinese).

Nathan, Andrew J., *Chinese Democracy.* New York: Alfred A. Knopf, 1985.

——, 'The Legislative Yuan Elections in Taiwan: Consequences of the Electoral System,' *Asian Survey,* 33 (April 1993): 424–38.

Niou, Emerson M. S. and Prdeshook, Peter C., 'A Game Theoretic Analysis of the Republic of China's Emerging Electoral System,' *International Political Science Review,* 13 (January 1992): 59–79.

Pang, Chien-kuo, *The State and Economic Transformation: The Taiwan Case*. New York: Garland Publishing, 1992.

Peng, Huy-en, *Toward High Status, the Cabinet Elites in the Republic of China*. Taipei: Dong-tsa Press, 1986 (in Chinese).

Peng, Huy-en, (ed.), *Taiwan's Party System in Political Development*. Taipei: Dong-tsa Press, 1987 (in Chinese).

———, *Formation of a Political Party by the Tang-Wai*. Taipei: Feng-Yun Lun-Tan Press, 1986 (in Chinese).

Peng, Ming-min, *A Taste of Freedom: Memoirs of a Formosan Independence Leader*. New York: Holt, Rinehart and Winston, 1972.

Severinghaus, Sheldon R., 'The Emergence of an Environmental Consciousness in Taiwan,' paper presented at the 1989 annual meeting of the Association for Asian Studies, Washington, DC.

Shaw, Yu-ming (ed.), *Building Democracy in the Republic of China*. Taipei: The Asia and World Institute, 1984.

Shen, Shin-yuan, *Kuomintang and Tang-wai: 1983 Legislative Yuan Election*. Taipei: Kuei-kuan Books, 1986 (in Chinese).

Shyu, Huo-yan, *Exploring Democratic Transition in Taiwan: An Analysis of Macro and Micro Political Changes*. Ph.D. Dissertation, Florida State University, 1990.

Sih, Paul K. T. (ed.), *Taiwan in Modern Times*. New York: St. John's University Press, 1973.

Soong, James Chul-yul, 'Political Development in the Republic of China, 1985–1992: An Insider's View,' *World Affairs*, 155 (Fall 1992): 62–6.

The Taiwan Issue, Proceedings of the Symposium, November 6, 1975, Asian Studies Center, Michigan State University.

Tien, Hung-mao, 'Origins and Development of Taiwan's Democratic Change,' paper presented at the 1989 annual meeting of the Association for Asian Studies, Washington, DC.

———, *The Great Transition: Political and Social Change in the Republic of China*. Stanford: Hoover Institution Press, 1989.

Tsai, Su-lin, 'Securing Social Status: the Comparison between Aborigines, Mainlanders, Taiwanese, and Hakka,' 1987, a paper for the Conference on Investigation on the Social Change in Taiwan Area in 1987 (in Chinese).

The United Daily News, a Chinese daily newspaper published in Taiwan.

Wade, Robert, *Governing the Market: Economic Theory and the Role of Government in East Asian Industrialization*. Princeton: Princeton University Press, 1990.

Wang, Shen, *A Research on The Three Principles of People*. Taipei: Central Cultural Supply, 1967 (in Chinese).

Wei, Yong, *Science, Talent, and Modernization*. 2nd edn. Taipei: Taiwan Student Press, 1985 (in Chinese).

Wilson, Richard, W., 'A Comparison of Political Attitudes of Taiwanese

Children and Mainlander Children on Taiwan,' *Asian Survey*, 8 (1968): 988–1000.

———, 'Some Rural-Urban Comparisons of Political Socialization in Taiwan,' *Asian Studies*, 10 (1972): 108–30.

Winckler, Edwin A., 'Institutionalization and Participation in Taiwan: From Hard to Soft Authoritarianism?' *China Quarterly*, 99 (September 1984): 481–99.

Yao, Chia-wen, *Democracy, Self-Determination, and Save Taiwan*. Taipei: Sen-huo wen-hua Press, 1988 (in Chinese).

Yen, Min-shen, *Three Principles of People and the Confucius Ethics*. Taipei: San-Min-Chu-Yi Research Institute, 1967 (in Chinese).

II. Other Publications

Almond, Gabriel A. and Verba, Sidney, *The Civic Culture: Political Attitude and Democracy in Five Nations, An Analytic Study*. Boston: Little, Brown, 1965.

Arat, Zehra F., 'Democracy and Economic Development: Modernization Theory Revisited,' *Comparative Politics*, 21 (October 1988): 21–36.

Avakian, Bob, *Democracy: Can't We Do Better Than That?* Chicago: Banner Press, 1986.

Bahrampour, Firouz, *Turkey: Political and Social Transformation*. Brooklyn: Theo. Gaus Sons, Inc., 1967.

Belloni, Frank and Beller, Dennis C. (eds.), *Faction Politics: Political Parties and Factionalism in Comparative Perspective*. Santa Barbara: ABC-Clio, 1978.

Bogdanor, Vernon and Butler, David (eds.), *Democracy and Elections: Electoral Systems and Their Political Consequences*. Cambridge, UK: Cambridge University Press, 1983.

Bollen, Kenneth A., 'Political Democracy: Conceptual and Measurement Traps,' *Studies in Comparative International Development*, 25 (Spring 1990): 7–24.

———, 'Political Democracy and the Timing of Development,' *American Sociological Review*, 44 (August 1979): 572–87.

Cnudde, Charles F. and Neubauer, Deane E. (eds.), *Empirical Democratic Theory*. Chicago: Markham Publishing Company, 1969.

Cohen, Carl, *Democracy*. New York: Free Press, 1971.

Coleman, James S., 'Democracy in Permanently Divided Systems,' *American Behavioral Scientist*, 25 (March/June 1992): 363–74.

Coppedge, Michael and Reinicke, Wolfgang H., 'Measuring Polyarchy,' *Studies in Comparative International Development*, 25 (Spring 1990): 51–72.

Coverdale, John F., *The Political Transformation of Spain After Franco*. New York: Praeger Publishers, 1979.

Cutright, Phillips, 'National Political Development: Measurement and Analysis,' in Cnudde, Charles F. and Neubauer, Deane E. (eds.), *Empirical Democratic Theory*. Chicago: Markham Publishing Company, 1969, 193–209. (Originally appeared in *American Sociological Review*, 28 [April 1963]: 253–64.)

Czudnowski, Moshe M. (eds.), *Does Who Governs Matter? Elite Circulation in Contemporary Societies*. Chicago: Northern Illinois University Press, 1982.

Dahl, Robert A., *Polyarchy: Participation and Opposition*. New Haven: Yale University Press, 1971.

———, *A Preface to Democratic Theory*. Chicago: The University of Chicago Press, 1956.

———, *Regimes and Opposition*. New Haven: Yale University Press, 1973.

Dahl, Robert A. (ed.), *Political Opposition in Western Democracies*. New Haven: Yale University Press, 1966.

Deutsch, Karl, 'Social Mobilization and Political Development,' *American Political Science Review*, 55 (September 1961): 493–514.

Diamond, Larry, *The Social Foundations of Democracy: The Case of Nigeria*. Ph.D. Dissertation, Stanford University, 1980.

———, 'Economic Development and Democracy Reconsidered,' *American Behavioral Scientist*, 35 (March/June 1992): 450–99.

Diamond, Larry; Linz, Juan J.; Lipset, Seymour Martin (eds.), *Democracy in Developing Countries*. 4 vols. Boulder: Lynne Rienner Publishers, 1989.

Eckstein, Harry and Gurr, Ted Robert, *Patterns of Authority: Structural Basis for Political Inquiry*. New York: John Wiley & Sons, 1975.

Evans, Peter B., *Dependent Development: The Alliance of Multinational, State and Local Capital in Brazil*. Princeton: Princeton University Press, 1979.

Evin, Ahmet (ed.), *Modern Turkey: Continuity and Change*. Opladen, Germany: Leske Verlag + Budrich GmbH, 1984.

Gastil, Ramond Duncan, 'The Comparative Survey of Freedom: Experiences and Suggestions,' *Studies in Comparative International Development*, 25 (Spring 1990): 25–50.

———, *Freedom in the World: Political Rights and Civil Liberties 1988–89*. New York: Freedom House, 1989.

Gendzier, Irene L., *Managing Political Change, Social Scientists and the Third World*. Boulder: Westview Press, 1985.

Gillespie, Charles G., 'Uruguay's Return to Democracy,' *Bulletin of Latin American Research*, 4 (1985): 99–107.

Grofman, Bernard and Lijphart, Arend (eds.), *Electoral Laws and Their Political Consequences*. New York: Agathon Press, 1986.

Gunther, Richard, 'Constitutional Change in Contemporary Spain,' in Banting, Keith G. and Simeon, Richard (eds.), *The Politics of*

Constitutional Change in Industrial Nations: Redesigning the State.
London: Macmillan, 1985.

Gunther, Richard and Blough, Roger A., 'Religious Conflict and Consensus in Spain: A Tale of Two Constitutions,' *World Affairs,* 144 (Spring 1981), 366–412.

Gunther, Richard; Giacomo Sani; and Goldie Shabad, *Spain After Franco, The Making of A Competitive Party System.* Berkeley: University of California Press, 1986.

Gurr, Ted Robert, *Why Men Rebel?* Princeton: Princeton University Press, 1970.

Gurr, Ted Robert; Jaggers, Keith; and Moore, Will H., 'The Transformation of the Western State: The Growth of Democracy, Autocracy, and State Power since 1800,' *Studies in Comparative International Development,* 25 (Spring 1990): 73–108.

Gurr, Ted Robert and Ruttengburg, Charles, *The Conditions of Civil Violence: First Tests of A Causal Model.* Princeton: Princeton University, Research Monograph no. 28, 1967.

Held, David, *Models of Democracy.* Cambridge, UK: Polity Press, 1987.

Herz, John H. (ed.), *From Dictatorship to Democracy: Coping With the Legacies of Authoritarianism and Totalitarianism.* Westport: Greenwood Press, 1982.

Higley, John and Gunther, Richard (eds.), *Elites and Democratic Consolidation in Latin America and Southern Europe.* Cambridge, UK: Cambridge University Press, 1992.

Hirschman, Albert L., *Exit, Voice, and Loyalty: Responses to Decline in Firms, Organizations, and States.* Cambridge, MA: Harvard University Press, 1970.

———, *The Passions and the Interests: Political Arguments for Capitalism Before Its Triumph.* Princeton: Princeton University Press, 1977.

———, *Essays in Trespassing, Economic to Politics and Beyond.* Cambridge, UK: Cambridge University Press, 1981.

Hudson, Michael C., *The Precarious Republic: Political Modernization in Lebanon.* New York: Random House, 1968.

Huntington, Samuel P., 'Will More Countries Become Democratic?' *Political Science Quarterly,* 99 (1984): 193–218.

———, *Political Order in Changing Societies.* New Haven: Yale University Press, 1968.

———, 'Political Development and Political Decay,' *World Politics* 17 (April 1965): 386–430.

———, and Nelson, Joan, *No Easy Choice: Political Participation in Developing Countries.* Cambridge, MA: Harvard University Press, 1982.

———, 'How Countries Democratize,' *Political Science Quarterly,* 106 ·(1991–1992): 579–616.

——, 'Democracy's Third Wave,' *Journal of Democracy*, 2 (Spring 1991): 12–34.

——, *The Third Wave: Democratization in the Late Twentieth Century*. Norman: University of Oklahoma Press, 1991.

Ike, Nobutaka, *A Theory of Japanese Democracy*. Boulder: Westview Press, 1978.

Johnson, Dale L., *Middle Classes in Dependent Countries*. Beverly Hills: Sage Publications, 1985.

Karl, Terry Linn, 'Dilemmas of Democratization in Latin America,' *Comparative Politics*, 23 (October 1990): 1–21.

Kautsky, John H., *The Political Consequences of Modernization*. New York: John Wiley and Sons, 1972.

LaPalombara, Joseph and Weiner, Myhron (eds.), *Political Parties and Political Development*. Princeton: Princeton University Press, 1966.

Lerner, Daniel, *The Passing of Traditional Society, Modernizing the Middle East*. 2nd edn. New York: Free Press, 1964.

Levine, Daniel H., 'Paradigm Lost: Dependence to Democracy,' *World Politics*, 40 (April 1988): 377–97.

Lichbach, Mark Irving, 'An Evaluation of "Does Economic Inequality Breed Political Conflict?" Studies,' *World Politics*, 41 (July 1989): 432–70.

Lijphart, Arend, *Democracy in Plural Societies: A Comparative Exploration*. New Haven: Yale University Press, 1977.

——, *Democracies: Patterns of Majoritarian and Consensus Government in Twenty-One Countries*. New Haven: Yale University Press, 1984.

——, 'The Political Consequences of Electoral Laws: 1945–85,' *American Political Science Review*, 84 (June 1990): 481–96.

——, and Grofman, Bernard (eds.), *Choosing an Electoral System: Issues and Alternatives*. New York: Praeger, 1984.

Linz, Juan J., 'Totalitarian and Authoritarian Regimes,' in Greenstein, Fred and Polsby, Nelson (eds.), *Handbook of Political Science*, Vol. 3, *Macropolitical Theory*. Reading, MA: Addison-Wesley Publishing Company, 1975.

Linz, Juan J. and Stepan, Alfred (eds.), *The Breakdown of Democratic Regimes*. Baltimore: The Johns Hopkins University Press, 1978.

Lipset, Seymour Martin, *Political Man, the Social Basis of Politics*. Baltimore: The Johns Hopkins University Press, 1981.

——, 'Some Social Requisites of Democracy: Economic Development and Political Legitimacy,' *American Political Science Review*, 53 (March 1959): 69–105.

——, *The First New Nation, the United States in Historical and Comparative Perspective*. New York: W. W. Norton, 1979.

Livingson, William S. (ed.), *A Prospect of Liberal Democracy*. Austin: The University of Texas Bicentennial Committee, 1979.

Lofland, John, *Protest: Studies of Collective Behavior and Social Movements*. New Brunswick, NJ: Transaction Books, 1985.

Marcus, George E. (ed.), *Elites and Ethnographic Issues*. Santa Fe: University of New Mexico Press, 1983.

Marks, Gary, 'Rational Sources of Chaos in Democratic Transition,' *American Behavioral Scientist*, 35 (March/June 1992): 397–421.

Margolis, Michael, *Viable Democracy*. London: The Macmillan Press, 1979.

May, John D., 'Democracy, Organization, Michels,' *American Political Science Review*, 69 (1965), 417–29.

McCrone, Donald J. and Cnudde, Charles F., 'Toward a Communication Theory of Democratic Political Development: A Causal Model,' *American Political Science Review*, 61 (1967): 72–9.

Moore, Barrington, Jr., *Injustice: The Social Bases of Obedience and Revolt*. New York: M. E. Sharpe Inc., 1978.

——, *Social Origins of Dictatorship and Democracy*. Boston: Beacon Press, 1966.

——, *Authority and Inequality under Capitalism and Socialism*. Oxford: Clarendon Press, 1987.

O'Donnell, Guillermo; Schmitter, Philippe C.; and Whitehead, Laurence (eds.), *Transitions from Authoritarian Rule*. 4 vols. Baltimore: The Johns Hopkins University Press, 1986.

Palmer, Monte, *Dilemmas of Political Development: An Introduction to the Politics of the Developing Areas*. Itasca, IL: F. E. Peacock Publishers, 1989.

Pourgerami, Abbas, *Development and Democracy in the Third World*. Boulder: Westview Press, 1991.

Powell, G. Bingham, *Contemporary Democracies: Participation, Stability, and Violence*. Cambridge, MA: Harvard University Press, 1982.

Prosterman, Roy L., *Surviving to 300, an Introduction to the Study of Lethal Conflict*. Belmont: Duxbury Press, 1972.

Putnam, Robert D., *The Comparative Study of Political Elites*. Englewood Cliffs: Prentice-Hall, Inc., 1976.

Ranney, Austin and Kendall, Willmoore, *Democracy and the American Party System*. New York: Harcourt, Brace & World, 1956.

Richardson, Bradley M., *The Political Culture of Japan*. Berkeley: University of California Press, 1974.

——, *Politics in Japan*. Boston: Little, Brown & Company, 1984.

Rustow, Dankwart A., 'Transitions to Democracy, Toward a Dynamic Model,' *Comparative Politics*, 2 (April 1970): 337–63.

Sartori, Giovanni, *Parties and Party System: A Framework of Analysis*. Cambridge, UK: Cambridge University Press, 1976.

——, *The Theory of Democracy Revisited*. Chatham: Chatham House Publishers, 1987.

Schmitter, Philippe C., 'The Consolidation of Democracy and

Representation of Social Groups,' *American Behavioral Scientist*, 35 (March/June 1992): 422–49.

Schumpter, Joseph, *Capitalism, Socialism, and Democracy*. New York: Harper & Row, 1942.

Searing, Donald D., 'The Comparative Study of Elite Socialization,' *Comparative Political Studies*, 1 (1969): 471–500.

Share, Donard, 'Transitions to Democracy and Transition Through Transaction,' *Comparative Political Studies*, 19 (January 1987): 525–48.

Taagepera, Rein and Shugart, Matthew S., *Seats and Votes: The Effects and Determinants of Electoral System*. New Haven: Yale University Press, 1989.

Verba, Sidney, 'Comparative Political Culture,' in Pye, Lucian and Verba, Sidney (eds.), *Political Culture and Political Development*. Princeton: Princeton University Press, 1965.

Ward, Robert E. and Sakamoto Yoshikazu (eds.), *Democratizing Japan, The Allied Occupation*. Honolulu: University of Hawaii Press, 1987.

Zannoni, Paolo, 'The Concept of Élite,' *European Journal of Political Research*, 6 (1978): 1–30.

Zimmermann, Ekkart, 'Political Violence and Other Strategies of Opposition Movement,' *Journal of International Affairs*, 40 (Winter/Spring 1987): 325–51.

Index